TECHNICAL
REPORT

T0302526

The Defense Acquisition Workforce

An Analysis of Personnel Trends Relevant to Policy, 1993–2006

Susan M. Gates, Edward G. Keating, Adria D. Jewell, Lindsay Daugherty, Bryan Tysinger, Albert A. Robbert, Ralph Masi

Prepared for the Office of the Secretary of Defense

Approved for public release; distribution unlimited

RAND NATIONAL DEFENSE RESEARCH INSTITUTE

The research described in this report was prepared for the Office of the Secretary of Defense (OSD). The research was conducted in the RAND National Defense Research Institute, a federally funded research and development center sponsored by the OSD, the Joint Staff, the Unified Combatant Commands, the Department of the Navy, the Marine Corps, the defense agencies, and the defense Intelligence Community under Contract W74V8H-06-C-0002.

Library of Congress Cataloging-in-Publication Data is available for this publication.

ISBN: 978-0-8330-4496-9

Published 2008 by the RAND Corporation
1776 Main Street, P.O. Box 2138, Santa Monica, CA 90407-2138
1200 South Hayes Street, Arlington, VA 22202-5050
4570 Fifth Avenue, Suite 600, Pittsburgh, PA 15213-2665
RAND URL: http://www.rand.org/
To order RAND documents or to obtain additional information, contact
Distribution Services: Telephone: (310) 451-7002;
Fax: (310) 451-6915; Email: order@rand.org

Preface

The defense acquisition workforce (AW), which includes more than 126,000 military and civilian personnel, is responsible for providing a wide range of acquisition, technology, and logistics support (products and services) to the nation's warfighters. The Under Secretary of Defense for Acquisition, Technology, and Logistics (USD (AT&L)) recognizes that the human capital represented by this workforce is a critical asset that must be strategically managed to support achieving successful Department of Defense (DoD) acquisition outcomes. The USD (AT&L) has made it a top priority to support DoD human capital strategies and has directed deployment of a comprehensive workforce analysis capability to support enterprise-wide and component assessments of the defense acquisition workforce. The Director, AT&L Human Capital Initiatives (HCI), who also serves as the President of the Defense Acquisition University (DAU), is responsible to the USD (AT&L) for department-wide strategic human capital management for the defense acquistion workforce within the Department of Defense.

This technical report summarizes workforce analyses that RAND has undertaken in support of AT&L HCI/DAU and the larger human capital strategic planning challenges it oversees. The report covers three areas of analysis: (1) the civilian acquisition workforce, (2) the careers of acquisition workforce senior executive service members, and (3) the military acquisition workforce and its implications for the larger workforce. Several data sources from the period 1992 to 2006 provide the basis for analysis.

The report will be of interest to officials responsible for acquisition workforce planning in DoD and other parts of DoD, workforce managers more generally, and members of the defense acquisition community. This research was sponsored by AT&L HCI/DAU and conducted within the Forces and Resources Policy Center of the RAND National Defense Research Institute, a federally funded research and development center sponsored by the Office of the Secretary of Defense, the Joint Staff, the Unified Combatant Commands, the Department of the Navy, the Marine Corps, the defense agencies and the defense Intelligence Community. For more information on RAND's Forces and Resources Policy Center, contact the Director, James Hosek. He can be reached by email at James_Hosek@rand.org; by phone at 310-393-0411, extension 7183; or by mail at the RAND Corporation, 1776 Main Street, Santa Monica, California 90407-2138. More information about RAND is available at www.rand.org.

Contents

Figures

Summary

In DoD, the Office of the Under Secretary of Defense for Acquisition, Technology, and Logistics (OUSD (AT&L)) is responsible for developing a human capital strategic plan for the defense acquisition workforce. The DoD share of the federal acquisition workforce is large in itself, consisting of more than 126,000 military and civilian personnel in over a dozen different services and agencies and representing several functional communities. While much of the strategic human capital planning in DoD occurs within individual services and agencies, DoD-wide visibility over such groups is still essential to effective management and planning for a workforce that cuts across multiple organizational boundaries.

OUSD (AT&L) responsibility for human capital strategic planning, in turn, rests with the Director, AT&L Human Capital Initiatives (HCI), who also serves as the President of the Defense Acquisition University (DAU). AT&L HCI/DAU has asked RAND to conduct workforce analysis in support of strategic human capital planning. Through an analysis of the acquisition workforce (both current snapshots and trends over time), RAND has assisted AT&L HCI/DAU in identifying and better understanding analytical challenges and in developing strategies for improving workforce management in the future. This report documents RAND's efforts.

Data Sources and Methods

The Defense Manpower Data Center (DMDC) maintains rich data sources on DoD's military and civilian workforces. Because of congressional reporting requirements, even more information is available about the acquisition workforce. However, these data are not readily usable for workforce analysis. To create analytical files, RAND assembled data from several sources and linked records across time and across data files.

DMDC provided RAND with annual civilian inventory and transaction file data covering the period October 1, 1991, to September 30, 2006. The inventory data provide annual demographic "snapshots" of each civilian employee—e.g., his or her grade, location, education level. The transaction data complement the inventory data by noting "transactions" that occur to workers between inventory snapshots. Attrition transactions were of central interest to us.

DMDC also provided RAND with acquisition workforce person file data covering fiscal year (FY) 1992 through FY 2006. These data identify both military and civilian personnel who are part of the acquisition workforce and provide additional information on these workers relevant to the acquisition community.

We also analyzed the military Work Experience file (WEX). The WEX contains information on anyone who has served in the U.S. military since 1975.

The Civilian Acquisition Workforce: Profile

The civilian acquisition workforce hit a low of 77,504 as of September 30, 1999, after the post–Cold War drawdown. It climbed steadily to 119,251 as of September 30, 2005, then declined slightly to 113,605 as of September 30, 2006. The civilian AW is better educated and more experienced than the DoD civilian workforce as a whole. The civilian AW has a disproportionate share of employees who are currently eligible to retire and who are nearing full retirement eligibility—one of the concerns giving rise to this research.

Mitigating the AW's retirement eligibility is that civilians in the AW voluntarily and involuntarily separate from DoD employment at very low rates. In addition, AW employees who become eligible to retire tend to remain employed by DoD for longer than is true of other retirement-eligible DoD civilian employees.

The Careers of Acquisition Workforce Senior Executive Service Members

There were 454 AW Senior Executive Service members (SESs) as of September 30, 2006. Most were long-time DoD civilian employees; 393 of the 454 had been DoD civilian employees on September 30, 1992.

Sixty-one of the 454 were already SESs on September 30, 1992; 307 of the 454 were GS- or GM-15s for at least one year between 1992 and 2005. Durations spent as a GS- or GM-15 varied widely. The time-in-grade of the civilian AW SES population is much more variable than that of military officers.

Of the 454 AW SESs, 398 did not work in multiple military services between 1992 and 2006. For civilian employees, the system appears to both value and reward depth of experience, not breadth. To the extent that the FY 2006 AW SESs did change services, their moves tended to be from the Department of the Navy toward the Office of the Secretary of Defense (OSD) or the Army.

While the norm has been for AW SESs to be long-time DoD civilian employees, there are some exceptions. Still, many so-called "surprising SESs" (those without significant prior civilian federal government experience) were previously high-ranking military officers instead.

The Military Acquisition Workforce and Its Implications for the Civilian Acquisition Workforce

In FY 2006, more than one-third (39.6 percent) of new AW civilian employees had prior military experience. This reflects a steady increase from just over 20 percent in FY 1993 and is consistent with observed increases in the fraction of all DoD civilian new hires who have prior military experience. Though military members represent a minority of the AW overall, they appear to be an important and growing source of future AW civilian leaders, especially in light of the undersized (drawdown-era) cohorts of DoD civilians.

Recommendations

Better definition and tracking of the acquisition workforce would improve workforce planning. Our analyses reveal that imposing a precise definition on the acquisition workforce is difficult, with numerous DoD civilians being recategorized into and out of the acquisition workforce each year. This report analyzes the extent to which these recategorizations reflect substantive changes in the nature of the job a person is performing versus administrative decisions on the part of the defense components to classify a particular organization as an acquisition organization or an individual as an acquisition worker. We define a recategorization to be *administrative* if, for the recategorized employee, his or her agency (e.g., military service), bureau (e.g., major command), functional occupational group, occupational series, and pay plan stayed the same. We find that by this definition, most recategorizations, both into and out of the AW, are administrative, not substantive.

A lack of clarity as to the definition of the workforce and how it varies across organizations is a barrier to effective management of the acquisition workforce and the development of recruitment, training, and retention policies. OSD should work together with the services to revise data collection policy guidance and should use that guidance to improve consistency of human capital planning across organizations. More precisely defining the workforce is central to this outcome.

More-detailed analysis of the current acquisition workforce and historical trends could yield additional insight. In this report, we provide only a few examples of the type of supply analysis that could more fully inform the management process. While considerable workforce analysis should occur at the service or major command level, there are some cross-cutting areas in which a DoD-wide perspective, including a comparison across services and agencies, might be particularly valuable. For example, a better understanding of the careers and performance of individuals in key leadership positions and the careers of acquisition workforce alumni might be particularly useful for both DoD and the services. In addition, an improved understanding of the careers of new civilian hires with prior military experience would improve DoD's civilian acquisition workforce management.

Workforce analysis is only one step in an overall strategic human capital planning effort. In this report, we have provided an overview of the DoD acquisition workforce and examples of the types of workforce supply analyses that can be supported by DoD data. Nevertheless, it is important to note that supply analysis is only one step in a strategic human capital planning effort. Such analyses must be combined with demand analyses. A better understanding of workload drivers for the acquisition workforce, as well as of the relationship between changes in the acquisition process and workload levels, would facilitate strategic human capital planning efforts for the acquisition workforce.

Acknowledgments

We thank Rodger Madison for programming assistance and Jim Hosek for comments on earlier drafts of this research. We appreciate the insights we received from Garry Shafovaloff and Frank Anderson of DAU. We also appreciate comments on earlier drafts of this report from Larry Lacy, from RAND colleague Frank Camm, and from Carla Tighe Murray of the Congressional Budget Office. Former RAND colleague Carl Dahlman, now in the Office of the Secretary of Defense, provided substantial input in early stages of this project. Margot Lynn of the DAU and Susan Pinciaro and Carolyn Willis of the Navy helped us understand characteristics of the data files used in our analyses. Portia Sullivan, Samantha Walker, and Terry McMillan of the Defense Manpower Data Center provided us with access to the data we needed.

Finally, we thank Donna White for helping to format the document and Miriam Polon for carefully editing the final copy.

Of course, the authors alone are responsible for errors that remain in the document.

Abbreviations

AT&L	Acquisition, Technology, and Logistics
AW	acquisition workforce
CSRS	Civil Service Retirement System
DAU	Defense Acquisition University
DAWIA	Defense Acquisition Workforce Improvement Act
DMDC	Defense Manpower Data Center
DoD	Department of Defense
DoN	Department of the Navy
FERS	Federal Employees Retirement System
FY	fiscal year
GED	General Educational Development Test
GS/GM	General Schedule/General Manager
HCI	Human Capital Initiatives
OSD	Office of the Secretary of Defense
SES	Senior Executive Service
SSN	Social Security Number
OUSD (AT&L)	Office of the Under Secretary of Defense for Acquisition, Technology, and Logistics
USD (AT&L)	Under Secretary of Defense for Acquisition, Technology, and Logistics
VSIP	Voluntary Separation Incentive Pay
WEX	Work Experience file (members of the military)
YOS	years of service

CHAPTER ONE

Introduction

The Department of Defense (DoD) recognizes that human capital is a critical organizational asset that must be managed in support of the overall DoD strategy. While workforce issues in general, and human capital strategic planning efforts in particular, are important throughout DoD, the Acquisition, Technology, and Logistics (AT&L) workforce has received special attention.[1]

Broadly speaking, the 2006 *Quadrennial Defense Review Report* (DoD, 2006) pointed to several reasons why human capital strategic planning is a critical need for DoD: (1) the need to manage workforces in response to the transformation from the industrial age to the information age, (2) rapid changes in technology and globalization and their effects on human resource planning, (3) the National Military Strategy, which calls for a force capable of decisive effects, integration of the Joint total force, and enhancement of institutional and individual agility, and (4) a mission-focused workforce that is responsive to changing demands.

In DoD, the Office of the Under Secretary of Defense for Acquisition, Technology, and Logistics (OUSD (AT&L)) is responsible for developing a human capital strategic plan for the *acquisition* workforce. The DoD portion of the federal acquisition workforce is large in itself and consists of more than 126,000 military and civilian employees, as well as a large, but highly variable, number of contractors (depending on mission requirements, funding, and short-term changes to those parameters). These individuals work in over a dozen different services and agencies and represent several functional communities. In developing a strategic human capital plan, a critical challenge facing DoD is thus how to segment the workforce for planning purposes. While much of the strategic human capital planning in DoD occurs within individual services and agencies, there are important workforce groups that cut across organizational boundaries. DoD-wide visibility over such groups is essential to effective management and planning.

OUSD (AT&L) responsibility for human capital strategic planning rests with the Director, AT&L Human Capital Initiatives (HCI), who also serves as the President of the Defense Acquisition University (DAU). AT&L HCI/DAU asked RAND, in turn, to conduct workforce analysis in support of strategic human capital planning. Through an analysis of the acquisition workforce (AW) (both current snapshots and trends over time), we have assisted AT&L HCI/DAU in better understanding current workforce challenges in developing strategies for improving workforce management in the future. This report documents RAND's efforts.

[1] See, for instance, DoD, Acquisition, Technology, and Logistics (2006) and U.S. Government Accountability Office (2002).

As we emphasize in this introduction, workforce analysis is only one element of the strategic human capital planning effort. Nevertheless, it is an important element and a useful starting point for this effort given the rich data available to DoD for this purpose. This report documents the first step in a broader research effort that will ultimately contribute to other aspects of strategic human capital planning as well.

The Defense Acquisition Workforce: Recent History

Acquisition processes in DoD were scrutinized closely in the mid-1980s following reports of the procurement of many items at unreasonably high cost, including the popular example of hammers costing several hundred dollars each (Fairhall, 1987). Reviews were instigated within DoD and from outside, most notably by the Packard Commission during the Reagan administration. The perception at the time was that the defense AW underperformed and was too large. Years of investigation and review of the performance and qualifications of the AW led to the Defense Acquisition Workforce Improvement Act (DAWIA) of 1990. The primary purpose of this legislation was to address concerns about workforce quality, in part by imposing requirements on the training of acquisition workers. Specifically, DAWIA required DoD to establish education and training standards (including creating DAU, a key source of education for acquisition workers), as well as formal career paths for the AW across all services. DoD subsequently issued several instructions that addressed the improvements specified by DAWIA.

The defense AW is defined by DoD as

> [T]he personnel component of the acquisition system. The acquisition workforce includes permanent civilian employees and military members who occupy acquisition positions, who are members of an Acquisition Corps, or who are in acquisition development programs. (DoD Instruction 5000.55, p. 20.)

The AW is responsible for planning, design, development, testing, contracting, production, introduction, acquisition logistics support, and disposal of systems, equipment, facilities, supplies, or services that are intended for use in, or support of, military missions. A key role of the AW is to provide oversight of the acquisition process. Because of the breadth of the work carried out, the AW spans organizational boundaries within DoD, including the Army, Navy, Marine Corps, Air Force, Defense Logistics Agency, Strategic Defense Initiative Organization, and Special Operations Command (DoD Instruction 5000.55).

DoD Instruction 5000.55 is the key policy document issued in response to DAWIA. It established 12 acquisition workforce career fields: Program Management; Communications–Computer Systems; Contracting; Purchasing; Industrial Property Management; Business, Cost Estimating, & Financial Management; Auditing; Quality Assurance; Manufacturing & Production; Acquisition Logistics; Systems Planning, Research, Development & Engineering; and Test & Evaluation Engineering. The instruction also describes career paths within these career fields. DoD Instruction 5000.66 provides guidance for managing AW positions and career development. It designates and identifies positions as part of the AW and describes requirements for workers in these positions to attain and maintain specific competencies through education, training, and experience. This instruction also presents guidelines for the selection and placement of personnel in acquisition positions and establishes performance metrics. In short, these instructions define career paths within the AW and establish a set of requirements for training and development of individuals within these career paths.

A popular term used in discussing the management of government agencies is "rightsizing"—having the optimal or appropriate number of employees to accomplish the required tasks. A crucial aspect of managing the size of the AW has been determining an accurate *count* of the number of individuals who are part of this workforce. Historically, the AW count has proved challenging. The Packard Commission established a counting methodology, called the acquisition organization workforce approach, which counts all personnel employed by the 22 DoD acquisition organizations, regardless of occupation (DoD, Office of the Inspector General, 2006). The refined Packard counting system is a revision of the acquisition organization counting system that excludes personnel who are not directly involved in acquisition support functions (e.g., human resources personnel or administrative assistants). Under the refined Packard system, services have varying levels of flexibility to count individuals as part of the AW depending on their occupation and the organization in which they work. The system defines three categories of personnel. Category I includes all DoD personnel in specified occupations, such as contracting, who are counted as part of the AW regardless of DoD organization. Category II includes personnel in other specified occupations, such as engineers and financial managers, who are counted only when serving in designated acquisition and technology organizations as discussed above. This category also includes all military officers assigned to designated acquisition organizations. Category III was added to give components flexibility to improve the accuracy of the count; it includes civilians, officers, and enlisted members performing acquisition functions but not categorized under I or II (Defense Acquisition University, 2007).

However, concern lingers over the validity of the reported counts, as counts from fiscal year (FY) 2004 and earlier are not verifiable (see DoD, Office of the Inspector General, 2006). Results of both of these counting methodologies are reported to Congress annually. Beginning with FY 2005, the DAWIA count has replaced the refined Packard count. According to the June 2007 *Defense Acquisition Structures and Capabilities Review*, this count is based directly on the incumbents' acquisition position responsibilities (Defense Acquisition University, 2007).

Establishing the size of the workforce is necessary, but doing so correctly is the subject of ongoing scrutiny.

Strategic Human Capital Planning for the Acquisition Workforce

The President's Management Agenda for FY 2002 presented itself as a "bold strategy for improving the management and performance of the federal government" (Executive Office of the President, 2002, p. 1). A key initiative in the agenda is the strategic management of human capital within the government agencies. A number of concerns were outlined in the agenda, including the approaching retirement of the baby-boomer generation, the weakness of human resources planning across government agencies, and the need for better recruiting, retention, and reward programs for workers. Toward this end, DoD generated a DoD-wide strategic human capital plan, and the USD (AT&L) developed a strategic human capital plan for the AW, which is currently in its third revision (see DoD, 2007). The existence of DAWIA and the DoD Instructions might position the AW well for meeting the President's goals because it has allowed for more than a decade's experience in human capital development. However, specific goals of the human capital strategic plan—in particular, the desire for a flexible workforce—might be encumbered by the rigidity of DAWIA's provisions.

Key issues raised in the AW human capital strategic plan include the following: the eventual loss of retirement-eligible personnel and their knowledge, understanding the differences in the workforce generations (aging baby-boomers compared with Generations X and Y, for example), and coping with the increasing demand for workers educated in science and engineering. The human capital strategic plan is summarized in these six goals:

1. Align and fully integrate with overarching DoD human capital initiatives.
2. Maintain a decentralized execution strategy that recognizes the components' lead role and responsibility for force planning and workforce management.
3. Establish a comprehensive, data-driven workforce analysis and decisionmaking capability.
4. Provide learning assets at the point of need to support mission-responsive human capital development.
5. Execute the DoD AT&L Workforce Communication Plan that is owned by all AT&L senior leaders.
6. Recruit, develop, and retain a mission-ready DoD AT&L workforce through comprehensive talent management.

The sixth goal was added in the third revision of the human capital strategic plan. These goals are clearly wide in scope, but they address key workforce issues: labor supply, labor demand, and human capital development, as well as meeting government-wide goals.

RAND has done several analyses of this multifaceted problem, typically at the service level. For example, a 2005 Arroyo Center study looked at acquisition reform issues in the Army acquisition processes, focusing primarily on the transformation of the 1990s (Hanks et al., 2005). Interviews of program managers during this study suggested ways to continue improving Army acquisition processes. Broadly, the action items suggested increasing access to education resources for personnel, changing Army policy to provide more support for acquisition workers attempting acquisition reform, and making resources available to personnel who are undertaking risks (such as new cost-cutting measures). Each of these points is related to the human capital strategic plan initiative through education or workforce development.

Workforce Analysis in Support of Human Capital Strategic Planning

Human capital strategic planning is a process that integrates organizational-level strategic planning with human resource planning and ties the latter to the strategic goals of the organization. It encompasses five broad tasks: (1) development of a strategic direction for the organization and the subsequent alignment of that direction with human resource development and succession, (2) workforce analysis (supply, demand, and gap analysis), (3) development of plans to address workforce gaps, (4) acquisition of resources to enable implementation, and (5) evaluation and revision of plans.

Workforce analysis is an essential part of strategic human capital planning. Typically, three major steps are involved in workforce analysis. Step 1 is to forecast demand, i.e., to estimate the staffing levels and competencies required in the future workforce. The term "workforce requirements" is often used to describe the output of the demand forecast. These requirements reflect the required number of positions and characteristics that the workers who fill those positions must have in order for the organization to meet its strategic intent. Employee characteristics that are measurable and potentially relevant to the identification of personnel requirements include skills or competencies, occupation/job series, and education. Ideally, an organization will have a model that translates expected workload into workforce requirements. The demand forecast should also identify the factors that affect workforce requirements and consider how those factors will change in the future. It should also consider the effect of technology on workforce demand.

Although much of the literature on strategic human capital planning emphasizes the importance of rigorous, data-driven demand analysis, as a practical matter it can be difficult, if not impossible, to obtain the data required for such analyses—particularly with regard to white-collar functions. Although there are certainly examples of data-driven demand analysis to be found in DoD, particularly among activities that have been slated for outsourcing and subject to a formal cost comparison study (see Gates, Eibner, and Keating, 2006, for specific examples) most workforce demand estimates or staffing decisions are informed primarily by historical experience and supervisory expertise.

Step 2 (which may be performed in tandem with Step 1) is to project workforce supply. This involves projecting current staffing levels and competency profiles into the future based on current trends in hiring, attrition, and retention. Step 3 brings together the results of Steps 1 and 2 to identify any gaps between supply and demand (recognizing that the supply estimates are a straightforward projection into the future that assumes no changes in workforce policy). This gap analysis may reveal important differences between the supply projection and demand forecast for particular organizational subunits, particular occupations, or specific competencies. Next, the strategic human capital planning process moves on to its third major element, addressing gaps: i.e., developing hiring, training, retention, and related strategies. The larger process then moves to funding these strategies, and finally to evaluating them and adjusting plans accordingly.

Gates, Eibner, and Keating (2006) and Vernez et al. (2007) emphasize that strategic human capital planning can occur at various levels of an organization and that organizational levels have different perspectives on and different roles to play in workforce planning.

A number of acquisition issues suggested by the acquisition workforce human capital strategic plan and/or RAND work require further research. These fit into five broad categories: labor supply, labor demand, developing the workforce, managing workforce separation, and steering labor supply to fit labor demand. Within labor supply are issues of the impending retirement of the baby-boomers, best practices for hiring new workers, and the best choice for workers in key leadership positions. Predicting future workforce requirements is the key aspect of labor demand and is an area ripe for additional study. The contracting workforce provides a potential source of talent for meeting labor demands, particularly because it adds flexibility to the workforce. Workforce separation issues include preserving the best workers and managing knowledge transfer between retirees and their replacements. Finally, using education resources in an optimal manner and structuring the AW for flexibility are both issues within workforce development.

Overview of RAND Acquisition Workforce Analysis

RAND has been working to assemble a comprehensive data file that can support a DoD-wide analysis of the DoD acquisition workforce—specifically the supply analysis described in Step 2 above. While such analysis is crucial to strategic human resource planning, it is only one component of strategic human capital planning efforts.

The RAND data file comprises information drawn from several files maintained by the Defense Manpower Data Center (DMDC): the DoD civilian personnel inventory file, the DoD civilian personnel transaction file, the military work experience file, and the acquisition workforce person file (5000.55 submission data). We have obtained data going back to FY 1992 for this work. Records can be linked across files (for example, between the military and civilian files or between the civilian inventory file and the acquisition workforce person file) and over time.

These DMDC files contain rich information on personnel including their position, assignment, rank, pay, occupation, years of service, demographic characteristics, education, acquisition career field, and acquisition certification level. By linking records across time and across files, we are able to examine movement into and out of the AW, as well as promotion and experience trajectories. In examining movement into and out of the AW, we are able to distinguish those who come from (or remain in) DoD from those who come from outside DoD (or who leave DoD altogether). The civilian transaction file provides us with information as to whether separation was voluntary or involuntary. We emphasize that no single DMDC database has all the information needed for the types of analyses that we are currently performing. For example, the acquisition workforce person file (5000.55 submission) includes only information on the AW, not on other civilian or military employees. Our analysis of acquisition turnover and specifically our ability to distinguish losses from DoD from switches into and out of the AW relies on our ability to link these data to data on the DoD civilian workforce as a whole.

Prior AW analyses were based on cross-sectional data. Although cross-sectional descriptive information can be examined over time, trends based on such cross-sectional information can obscure information that is highly relevant for workforce planning. Moreover, cross-sectional analyses cannot support an analysis of career trajectories.

Outline of Report

The rest of this report is structured as follows. Chapter Two discusses the data sources we used in our analysis. Chapter Three then presents factual information about the civilian AW. Chapter Four provides a detailed example of the way in which data on the civilian workforce can be used by managers. The chapter describes, in depth, the inventory projection model that uses data on the civilian AW as a key input. Readers who are not interested in workforce projections may wish to skip this chapter. Chapter Five focuses on Senior Executive Service members (SESs) in the AW and their careers. Chapter Six discusses the military AW and military experience of new civilian AW hires. Chapter Seven presents our conclusions and recommendations.

CHAPTER TWO

Data Sources and Methods

To undertake the analyses presented in this report, RAND brought together several data sources. The Defense Manpower Data Center provided us with annual civilian inventory and transaction file data, data from the acquisition workforce person file, and military work experience file data. The coverage of each file varies, but for each file we have data covering October 1, 1991, to September 30, 2006. The data provided to RAND included scrambled Social Security Numbers (SSNs). The same scrambling algorithm was used for each year and for each data file. This enabled us to link individual records over time and merge information across files. Among other things, this allows us to track individuals as they move into and out of the AW and between the military and civilian workforces in DoD.

The civilian inventory data provide annual "snapshots" of each civilian employee, e.g., their grade, location, education level, and other demographic variables as of September 30. The data also include information on an individual's occupation, the organization he or she works in, and the individual's pay plan and years of service.

The transaction data complement the inventory data by noting "transactions" that occur to workers between inventory snapshots. The transactions of central interest to us were indicators of attrition, e.g., retirement, voluntary separation, involuntary separation. In theory, if we see a worker in, say, the September 30, 1997, inventory but we do not see him or her in the September 30, 1998, inventory, we expect to see a transaction between those dates that notes his or her departure. Unfortunately, the data are not always as interpretable as that. Sometimes we do not find a departure transaction or sometimes we find it prior to the last inventory date. These are administrative data that have imperfections. But, in general, the combination of inventory and transaction files presents a reasonably coherent portrayal of the evolution of the civilian workforce.

DMDC also provided RAND with access to the military Work Experience file (WEX). The WEX contains information (e.g., scrambled SSN, rank, military service) on anyone who has served in the U.S. military since 1975. Hence, we were able to check whether any specific DoD civilian employee had served in the military since 1975 and, if so, what rank he or she achieved and in which military service(s) he or she served.

DMDC also provided us with the acquisition workforce person file for FY 1992 through 2006. This file contains a record for each individual (both military and civilian) who was included in the service or agency submissions made in accordance with DoD Instruction 5000.55.[1] By matching the records in the 5000.55 file with the records in the civilian inven-

[1] There have been a number of controversies about how the AW is tallied. See DoD, Office of the Inspector General (2006). We use the 5000.55 tabulation throughout this analysis. Our explorations of other approaches, e.g., the so-called "Refined Packard" methodology, had findings similar to those presented in this report.

tory file and the WEX file, we were able to create a variable in the annual inventory snapshots indicating whether a specific individual was or was not in the AW that year.

Data Limitations

The data we use for this analysis were designed for administrative, not research, purposes. The data are extraordinarily comprehensive, providing information on every DoD military and civilian employee. We do not, however, have information on contractor employees. There are also some limitations that are specific to the data files that we are using. Since our WEX data go back only to 1975, we cannot observe Vietnam era military experience. However, even by 1992, it seems unlikely that many new hires had previous military experience that entirely predated 1975. We generally restrict our analysis of the military experience of civilian acquisition workers to new hires. The acquisition workforce person files provided to us by DMDC contain information on both "incumbents" and "nonincumbents." Apparently, in submitting their AW counts to DoD, the services can "count" individuals who may be part of the AW (i.e., in the acquisition career field) but are not currently in acquisition positions. These people are referred to as nonincumbents. Based on analysis of data obtained from the Navy, we find that about 1 percent of the AW as reported in the DoD-wide data are nonincumbents.

Methods

The analyses that we present in this report are descriptive in nature. Unless otherwise noted, the information on the civilian AW and on all DoD civilians includes all appropriated-fund civil service employees. We did not exclude civilians from the analysis based on their pay plan or because they are employed part-time. For some analyses, we do restrict the population under consideration based on their education level or their pay plan. Our analysis of military personnel includes active-duty military members, including activated reservists.[2]

Analysis of Attrition

In analyzing attrition from one year to the next, we look for people who are in the data set in one year but not the next year. If a person "leaves" the data set, we then check to see whether the transaction file includes a departure transaction for that individual. The civilian transaction file includes scores of different departure codes that describe why a person has left DoD. We group those departure codes into the following categories: retirement, voluntary separation, involuntary separation, death, and other. Although the distinctions are usually obvious, some of the categorizations require judgment calls. For example, we code "retirement in lieu of involuntary action" as an involuntary separation. If the next year's transaction file—or, in some cases, the current year's transaction or inventory file—includes a departure transaction, we designate the individual a "leaver" and assign to the record the appropriate attrition code based on the categories described above. Sometimes, an individual leaves the data set but has

[2] We track the military experience of nonactivated reservists but do not include them as part of the military workforce count. Many of these reservists are currently serving as DoD civilian employees. Future work will strive to better understand the role of nonactivated reservists in the AW.

no separation code in the transaction file. Our review of historical data suggests that many of these people reappear in the data set in future years. In these instances, we look forward in the data as much as possible, through September 30, 2006. When people reappear in a future year, they are no longer designated as "leavers" in the current year. Alternatively, it is possible for an individual never to reappear in future years. Since a separation code was never located for this person, the individual must be coded as "missing."[3]

Analysis of New Hires

Paralleling our analysis of attrition, we also analyze new hires. We designate individuals as new hires to DoD when they appear in the DoD civilian inventory file for the first time. We check the scrambled SSN of each new hire against the WEX file in order to determine whether the individual has prior military experience. If there is a match, we pull some basic descriptive information about the individual from the military file. In this way, we are able to analyze the military characteristics of new civilian hires.

Analysis of Recategorization

We identify individuals as part of the AW if they were flagged as such in the acquisition workforce person file for that fiscal year (DoD Instruction 5000.55 reports).

Because the AW is the focus of our analysis, we are often interested in looking at the population of people who have entered or left the AW in a given year. However, the AW is a subset of a larger population of DoD employees, so some but not all workers who are new to the AW are "new hires" into DoD (as described above). Others are people who were part of DoD in the past year, but not part of the AW.

We identify individuals who move between the DoD civilian AW and non-AW as *recategorizations.* We distinguish recategorizations from new hires into DoD and separations from DoD.

A recategorization may be administrative or substantive. We define a recategorization to be *administrative* if, for the recategorized employee, his or her agency (e.g., military service), bureau (e.g., major command), functional occupational group, occupational series, and pay plan stayed the same. This would suggest that the person's job was recategorized although nothing substantive changed in terms of the work being done.

This definition is conservative and likely *underestimate*s the fraction of recategorizations that are administrative. If any one of the categories mentioned above changes for any reason, the recategorization would not be classified as administrative. If, for example, an employee was reclassified and moved into a demonstration project pay plan in the same year, we would not identify that reclassification as administrative because "pay plan" changed.

Retirement Eligibility

We calculate the number of years relative to regular retirement eligibility for each civilian in the inventory file. Retirement eligibility depends on an individual's retirement plan, age, and years of service (YOS). Our measure does not account for special retirement incentives, early retirement options, or disability retirement. For this reason, we do observe some people in

[3] An important exception pertains to individuals who leave the data set, have no separation code, but do have a separation incentive code (825). These individuals are coded as voluntary separations or retirements based on their retirement eligibility at the time they leave the data set.

the data set who retire before having reached regular retirement eligibility. In any given year, individuals with zero years relative to retirement eligibility are those who became retirement-eligible for the first time in that year. Those with a negative number of years relative to retirement eligibility have not yet reached retirement eligibility (assuming that they continue to be employed full-time); those with positive years relative to retirement eligibility have already attained retirement eligibility.

In the next chapter, we present some basic facts about the civilian AW.

CHAPTER THREE

The Civilian Acquisition Workforce

In this chapter we present insights on the civilian AW gleaned from our analysis of DoD data. We first provide a descriptive overview of the civilian AW. We next discuss the issue of recategorizations into and out of the AW and some of the challenges that such recategorizations pose for workforce analysis. The final section presents findings from analysis of AW attrition. In Chapter Four, we present an example of how the descriptive information and data on attrition can be used by workforce managers to conduct workforce supply analysis.

Descriptive Overview

Figure 3.1 displays the civilian AW end-of-FY totals as tallied from DoD Instruction 5000.55–required service-provided reports. The civilian AW hit a low of 77,504 as of September 30, 1999, climbed steadily to 119,251 as of September 30, 2005, and then was reduced slightly to 113,605 by September 30, 2006. The defense AW emerged as a meaningful concept in the early 1990s with the enactment of the Defense Acquisition Workforce Improvement Act of 1990. The act required DoD to implement career development programs for this workforce. Throughout the 1990s, DoD worked to develop acquisition corps, establish certification standards for acquisition workers, and improve training and professional development (Garcia et al., 1997). Compliance with the act required DoD to begin counting and tracking the individuals considered to be part of the AW. The act also requires DoD to ensure that members of the AW receive high-quality training and professional development. It is not clear whether changes in the number of employees designated by DoD as part of the AW reflect real variation in the number of individuals performing actual acquisition functions or changes in the acquisition workload. As we discuss later in this section, changes in the size of the AW may result from changes in the way this workforce is defined.

Looking at Figure 3.2, we see that 38.7 percent of AW civilians were employed by the Army, 31.6 percent by the Navy, 13.9 percent by the Air Force, and 15.8 percent by the Office of Secretary of Defense (OSD) and the defense agencies in 2006.

As shown in Figure 3.3, the AW has significantly more education, particularly at the bachelor's and master's levels, than does the overall DoD civilian workforce.

Figure 3.4 shows that AW civilians are more experienced than is typical DoD-wide. There is a dearth of civilian workers with 5–14 years of experience, both in the AW and DoD-wide. This trough was caused by the post–Cold War DoD drawdown that resulted in limited civilian hiring in the 1990s.

Figure 3.1
Civilians in the Acquisition Workforce, September 30 Annual Snapshots

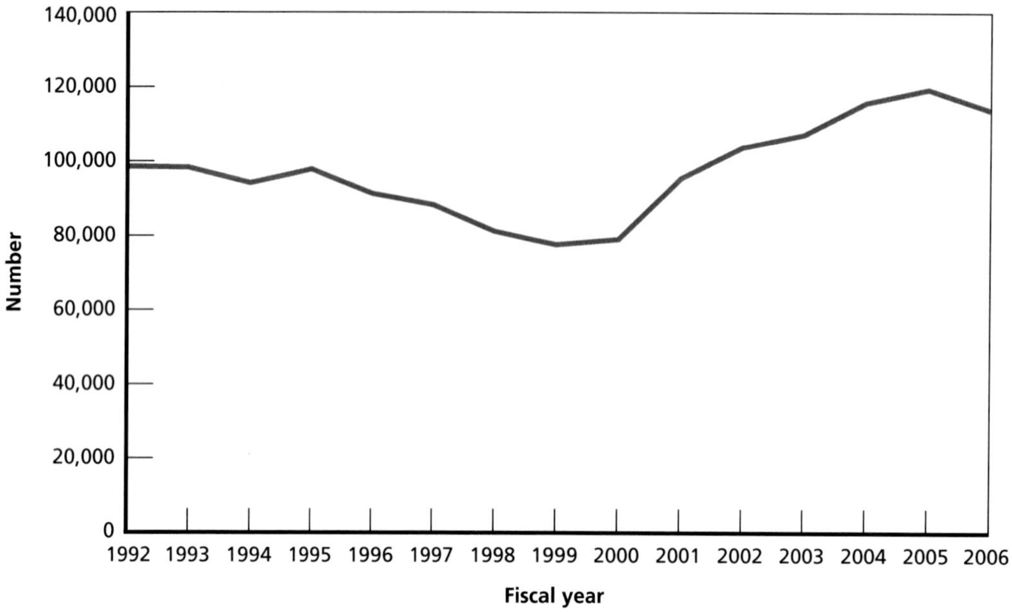

RAND *TR572-3.1*

Figure 3.2
AW Civilian Inventory, by Service or Agency, 2006

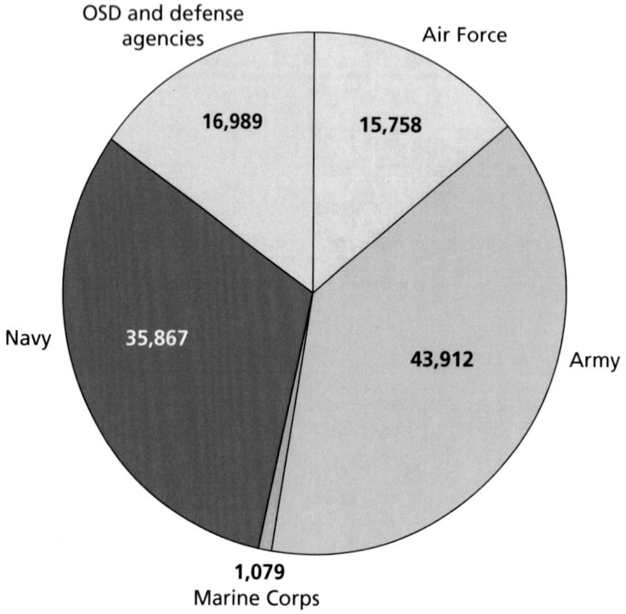

RAND *TR572-3.2*

Figure 3.3
AW and DoD Civilian Inventory Education Levels, 2006

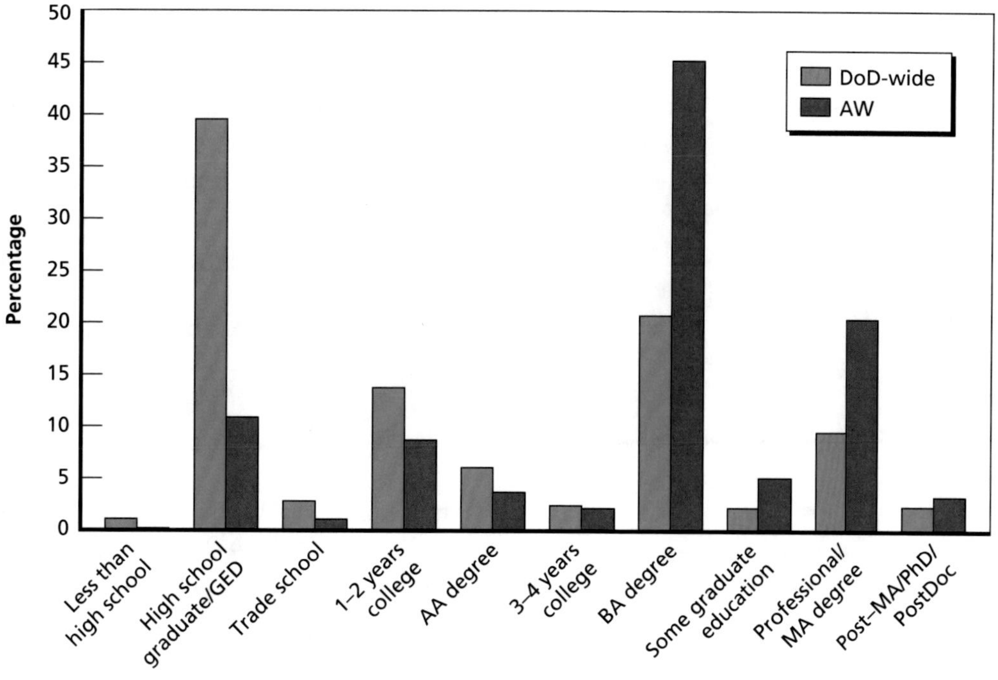

RAND *TR572-3.3*

Figure 3.4
AW and DoD Civilian Years of Service Levels, 2006

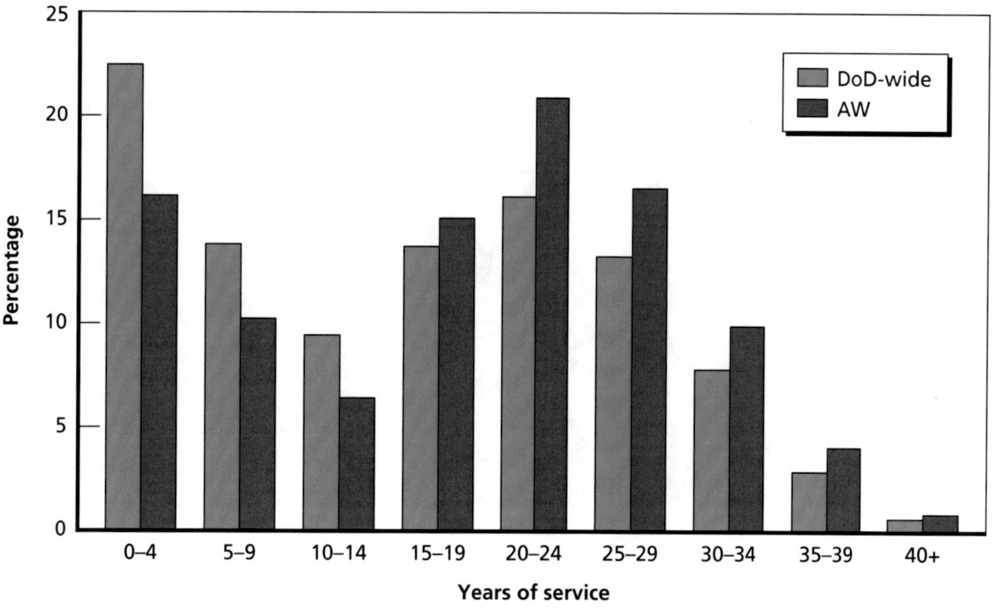

RAND *TR572-3.4*

The AW differs dramatically from the DoD civilian workforce as a whole in terms of pay plans. Whereas over 20 percent of all DoD civilians are in blue-collar pay plans (WG, WL, WS), almost no acquisition workers fall into such pay plans. Conversely, over 23 percent of DoD's AW falls under professional and administrative demonstration pay plans, whereas only 6 percent of all DoD civilians are in professional and administration pay plans.

It is also useful to consider the activities performed by the civilian AW as reflected by acquisition career fields. Figure 3.5 illustrates that the majority of DoD civilian acquisition workers work in one of two career fields: Systems Planning, R&D, Engineering; or Contracting. It is notable that only 7 percent of civilian acquisition employees are in Program Management. As we will see in Chapter Six, the distribution of military AW across career fields looks quite different.

Figure 3.6 shows the September 30, 2006, fraction of the total civilian workforce by years relative to retirement eligibility. In this display, a person at Year 0 became retirement-eligible during FY 2006. A person at Year 1 became retirement-eligible during FY 2005 (and chose to remain employed by DoD). A person at Year –1 will first become retirement-eligible in FY 2007 but was not retirement-eligible as of September 30, 2006.

In the coming years, about 4 percent of both the DoD and AW civilian workforce will be achieving retirement eligibility each year. Notice, however, that the AW has a looming peak of retirement-eligible civilians coming within the next decade or so—a major consideration in strategic human resource planning. Also, the AW has proportionally more employees who are already retirement-eligible but who have not chosen to leave DoD employment heretofore. It is this disproportionate degree of current and pending retirement eligibility that has motivated particular interest in AW personnel management issues.

Figure 3.5
Civilian AW, by Career Field, 2006

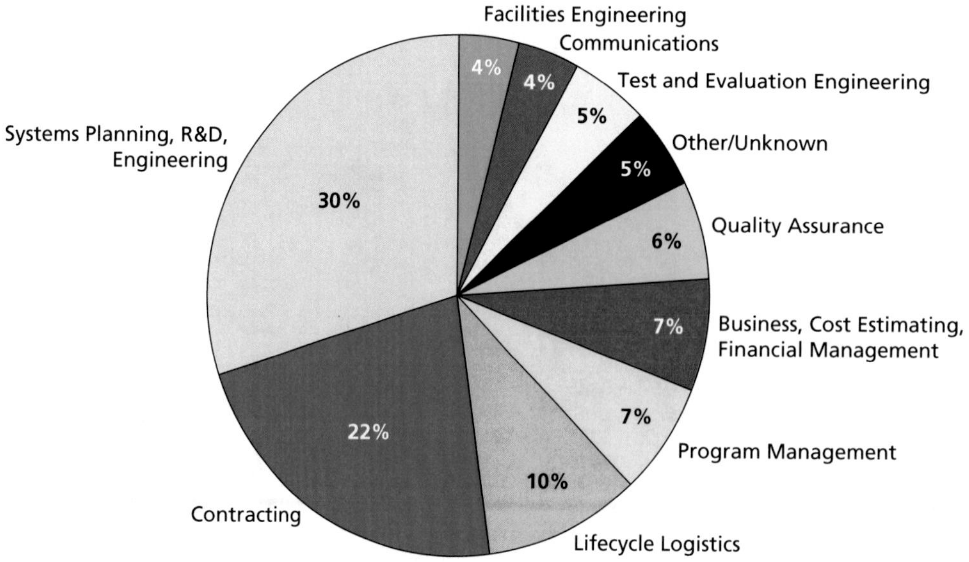

RAND *TR572-3.5*

Figure 3.6
Percentage of Civilian Workers with or Nearing Retirement Eligibility, September 2006

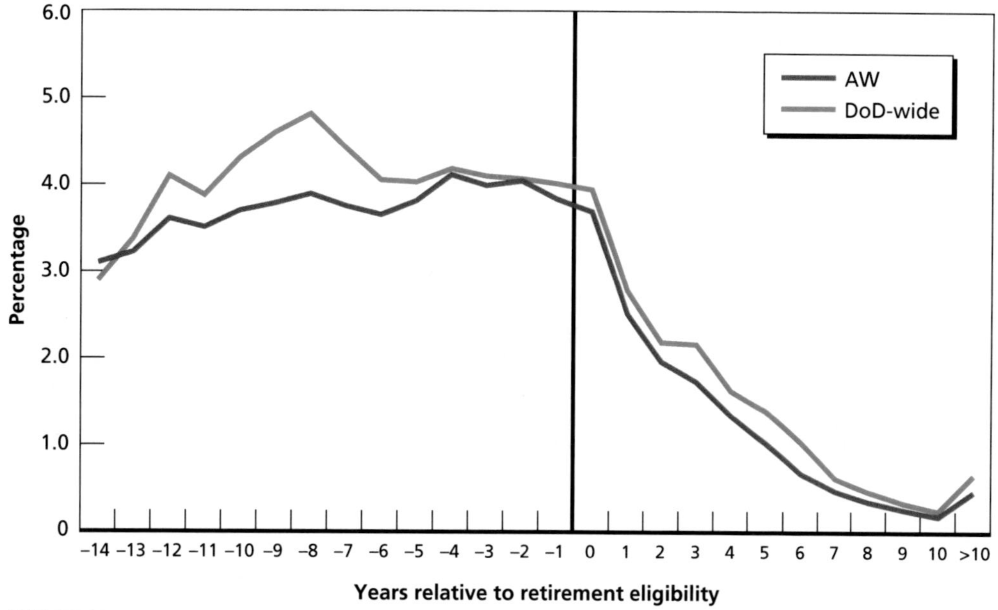

RAND *TR572-3.6*

Acquisition Workforce Recategorizations

Figure 3.1 shows fairly marked growth in the size of the civilian AW in recent years. Figure 3.7 plots the number of new hires entering the AW, civilians already employed by DoD who were recategorized into the AW, recategorizations out of the AW, and attrition out of DoD among members of the civilian AW by fiscal year.

Every fiscal year since 1993, more existing DoD employees were recategorized into the AW than there were new DoD employees hired into the AW. The number of new hires has increased in recent years, but hiring new DoD employees remains the less-common accession path into the AW.

On the other hand, recategorizations out of the AW (while remaining employed by DoD) and attrition out of DoD have been of generally comparable magnitudes.

In Figure 3.8, we break up the annual recategorization spikes by military service. The large 2001 spike into the AW was predominantly from the Army (15,287 of 20,513). The 2002 spike into the AW was predominantly from the Department of the Navy (DoN)—i.e, the Navy and the Marine Corps (8,117 of 15,247). FY 2006 saw an increase in the overall number of recategorizations out of the AW due primarily to an increase in such outbound recategorizations by the Air Force and the Army.

Consistent with observing large, service-concentrated spikes, we interpret many recategorizations as administrative, not substantive. In most cases, as best we can tell from these data, recategorized workers were performing the same (or very similar) tasks both before and after being put into or removed from the AW.

Figure 3.7
Entrances into and Exits from the Civilian Acquisition Workforce

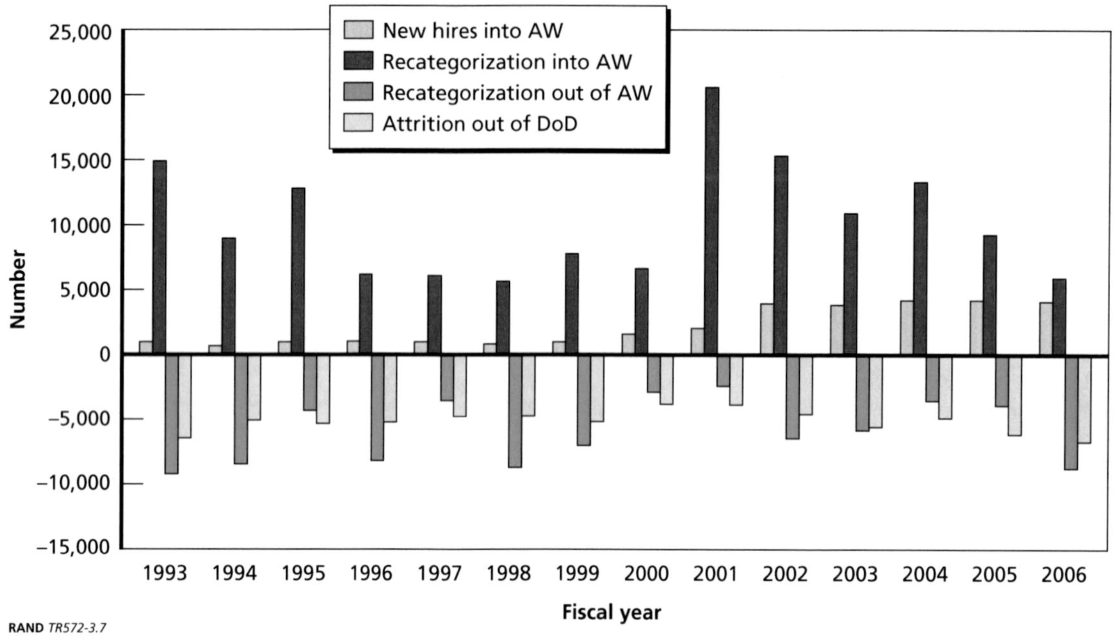

RAND *TR572-3.7*

Figure 3.8
Civilian Recategorizations, by Military Service

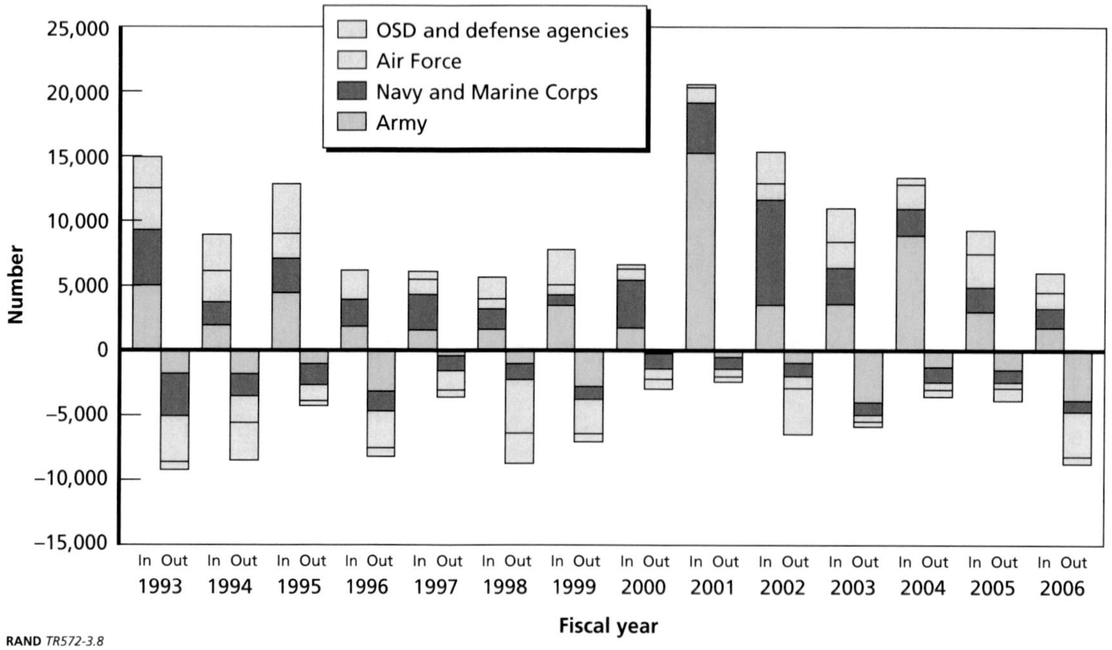

RAND *TR572-3.8*

In Chapter Two, we defined a recategorization to be administrative if, for the recategorized employee, his or her agency (e.g., military service), bureau (e.g., major command), functional occupational group, occupational series, and pay plan stayed the same. We believe this definition is broad and likely underestimates the share of recategorizations that are administrative. Using this definition, Figure 3.9 shows that most recategorizations, both into and out of the AW, are administrative.

Figure 3.9 suggests there has been more turnover in the AW's composition than actual changes in what DoD civilian workers have been doing. While membership in the AW might be presumed to be a stable construct, in fact, significant numbers of workers may have had their categorizations changed, perhaps unbeknownst to them. Indeed, it is not clear that a given worker always knows whether he or she is categorized as being in the AW.

Another possibility, of course, is that there are substantive aspects of workers' jobs that change upon entrance into or exit from the AW, but these aspects are not observed in the civilian personnel data. We have no way to evaluate this alternative possibility.

Acquisition Workforce Attrition

As shown in Figure 3.10, AW attrition out of the DoD has been consistently lower than DoD civilian norms, largely due to lower voluntary and involuntary separations. However, retirement rates (as a share of the total civilian workforce) have been comparable for AW civilians and all DoD civilians.

One reason AW attrition may be lower than DoD norms is that, as shown in Figure 3.3, AW workers tend to be better educated, and better-educated workers, in general, attrit at lower rates. Also, as shown in Figure 3.4, AW workers tend to have more years of service as DoD-employed civilians, and seniority correlates with lower departure probabilities. We explore the possibility that education and experience are key factors driving the lower rates of separation

Figure 3.9
Percentage of Recategorizations RAND Believes Were Administrative

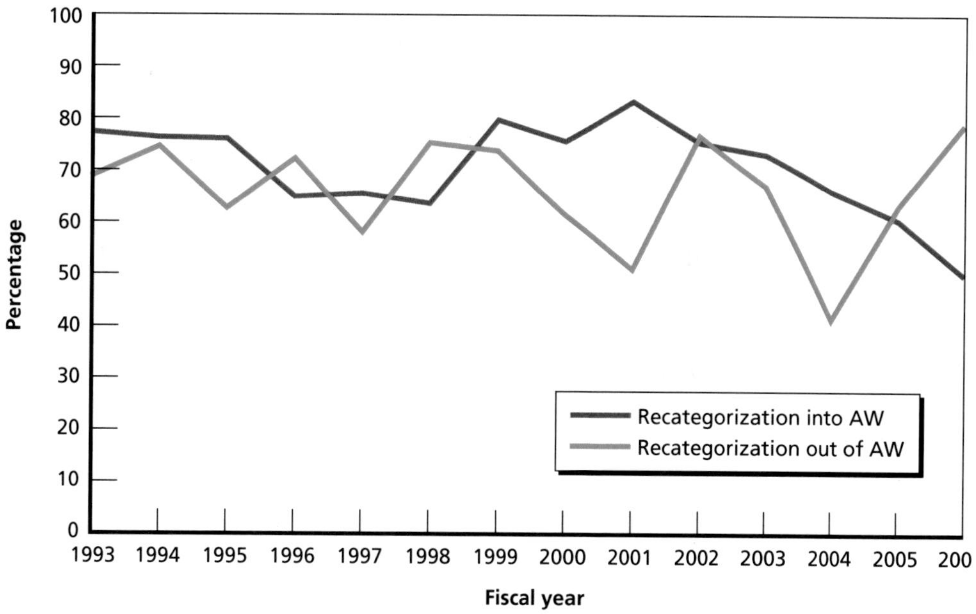

RAND *TR572-3.9*

Figure 3.10
DoD and AW Annual Civilian Attrition Rates

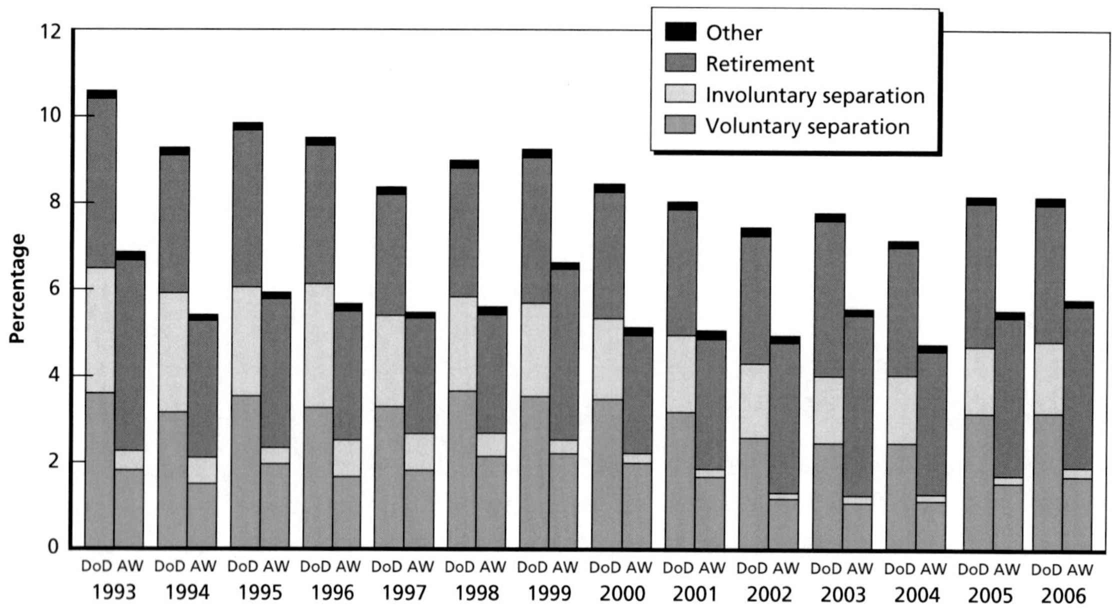

RAND *TR572-3.10*

among members of the AW. It is also worth noting that in FY 2006, 30 percent of retirement involved a separation incentive—in other words, DoD was actually encouraging people to retire.[1]

In Figure 3.11, we present information on rates of voluntary and involuntary separation among civilian employees with a BA degree as the highest education attainment by years of service. As workers become more senior, they are less likely to separate voluntarily or involuntarily. Note, however, that even within a specific YOS band—e.g., 10–14, AW voluntary and involuntary separation rates are lower than are found DoD-wide.

The AW seems to be characterized by lower attrition, even controlling for the different seniority and educational composition of the AW. Acquisition workers, simply put, seem unusually attached to their jobs relative to other DoD civilian employees.

Not surprisingly, though, a surge in attrition occurs when civilian employees become fully retirement-eligible, both DoD-wide and in the AW. Figure 3.12 uses the same x-axis as shown in Figure 3.6, i.e., the number of years relative to regular retirement eligibility for DoD civilians. Workers with zero years relative to retirement eligibility are those who became retirement-eligible for the first time in FY 2006. The far left side of the graph reflects those with a decade or more until retirement eligibility, while the far right reflects those who have been retirement-eligible for a decade or more but still remain employed by DoD. We have placed the vertical axis at the edge of Year 0, the first year of full retirement eligibility.

Along with the manifest jump in attrition propensity upon becoming fully retirement-eligible, we also see that the AW (again) has lower attrition in the years preceding eligibility compared with the DoD workforce as a whole. Also, in the years immediately after workers

---

[1] Our analysis of the data shows that 30 percent of those who retired also had a transaction file record indicating that they received a special separation incentive.

Figure 3.11
Voluntary and Involuntary Separation Rates for Civilian Employees with Bachelor's Degrees, 2005–2006

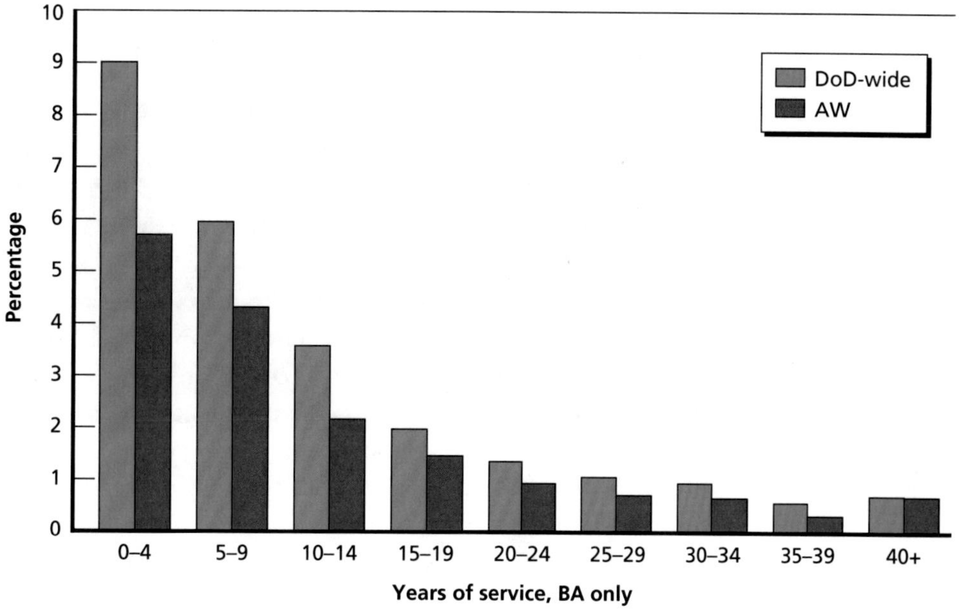

RAND *TR572-3.11*

Figure 3.12
Annual Attrition Rates Relative to Retirement Eligibility

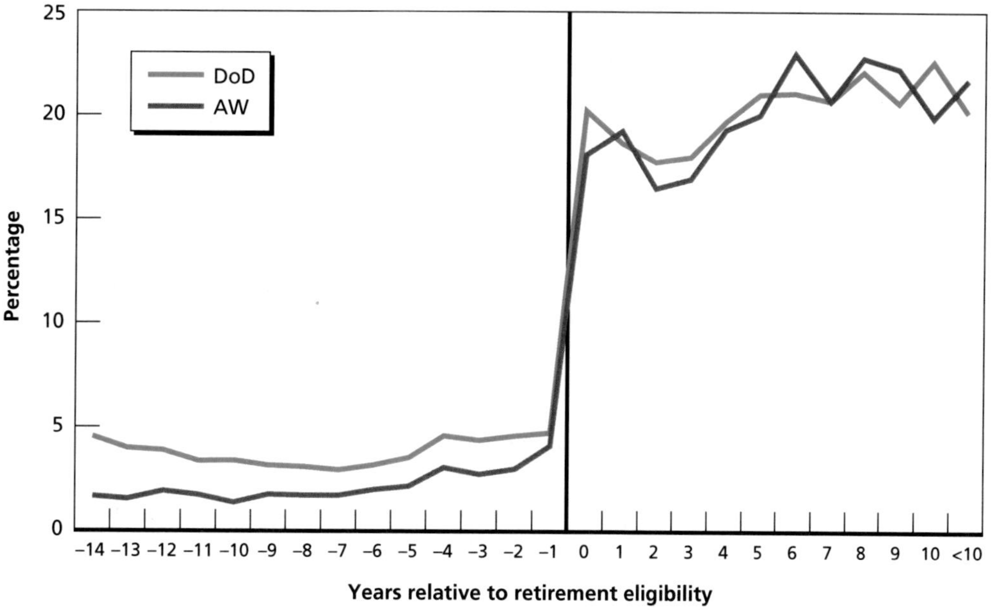

RAND *TR572-3.12*

have become retirement-eligible, AW employees typically do not depart as frequently as their non-AW colleagues. (Sample sizes are very small on the right side of the graph, e.g., employees who have been retirement-eligible for nine years, so we caution against overinterpreting the rightmost points in this figure.)

Civil servants currently employed by DoD participate in two main retirement plans: the Civil Service Retirement System (CSRS) and the Federal Employees Retirement System (FERS). FERS was created in 1986; anyone hired into the federal civil service after January 1, 1987, is automatically covered under FERS. Employees hired prior to that date were covered by CSRS when they were hired but had the option to switch into FERS. CSRS is a traditional defined benefit plan that provides those who attain specific retirement eligibility criteria with a retirement benefit in the form of an annuity. Those covered by CSRS are not eligible for Social Security benefits based on their federal employment. FERS has a defined benefit and a defined contribution component. In addition, individuals covered under FERS also receive Social Security credits. Under CSRS, employees who leave federal employment before they reach retirement age receive no retirement benefits. Thus, the plan creates very strong incentives for employees to remain in the civil service. These incentives do not exist with FERS to the same extent, since all employees receive a government contribution to their Thrift Savings Account (similar to a 401K account) and employees with as few as five years of service are eligible for a basic benefit annuity payment when they reach retirement age (U.S. Office of Personnel Management, 1997).

Given the differences between the two major retirement plans, it is important to consider whether the retirement behavior described in this section differs for individuals depending on their retirement plan.

Figure 3.13 illustrates why this is a concern. This figure shows the percentage of 2006 civilians covered by CSRS (as opposed to FERS or other retirement programs) as a function of years until retirement eligibility.

While most newer DoD-employed civilians are in FERS, many more senior employees are covered by CSRS. Currently, a majority of civilians reaching retirement eligibility are covered by CSRS, but that will change over the next decade. Figure 3.14 shows the attrition rate by years relative to retirement eligibility for CSRS and FERS employees in the AW during FY 2006.

Those AW civilians in CSRS generally have lower attrition rates before retirement eligibility but then a bigger attrition jump upon achieving eligibility relative to FERS AW civilians. This result is not surprising. Both CSRS and FERS reward attainment of retirement eligibility, but CSRS has no retirement benefits for those who depart early (outside of special early retirement programs). Hence, we expect (and observe) more CSRS employees "hanging on" to achieve full retirement eligibility, then leaving upon its attainment.

Summarizing, this section has provided an overview of the AW. We find that, compared with DoD civilians more generally, the AW has relatively low turnover up to retirement eligibility. Paradoxically, though, AW issues garner much attention—given this population's high level of attachment to DoD employment.

In the next chapter we provide an example of the way in which managers can use a simple workforce inventory projection model, which incorporates this descriptive information (including information distinguishing administrative from substantive recategorizations), for workforce planning purposes.

Figure 3.13
Percentage of FY 2006 DoD Workers Enrolled in CSRS

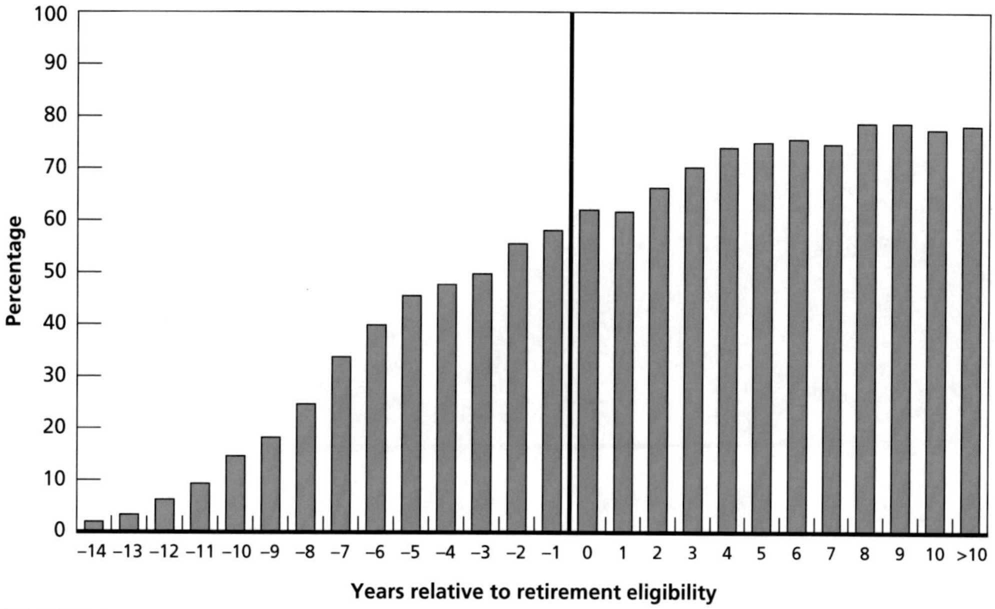

RAND *TR572-3.13*

Figure 3.14
AW Attrition Rate, by Retirement Plan, FY 2006

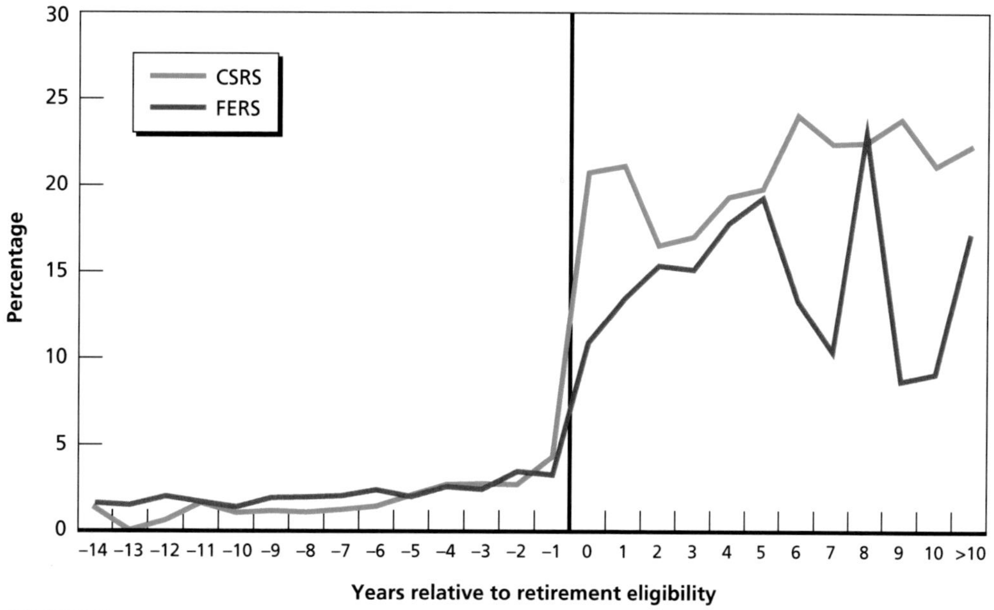

RAND *TR572-3.14*

CHAPTER FOUR

An Application: Acquisition Workforce Inventory Projections

Our workforce inventory projection model, shown in Figure 4.1, takes as its inputs information on the number of AW employees, rates of accession, rates of separation, and rates of recategorization, as described in Chapter Three. In our application of the model, workforce counts and turnover rates are calculated for each "years of service" group.[1]

Figure 4.1
Overview of the Inventory Projection Model Calculation

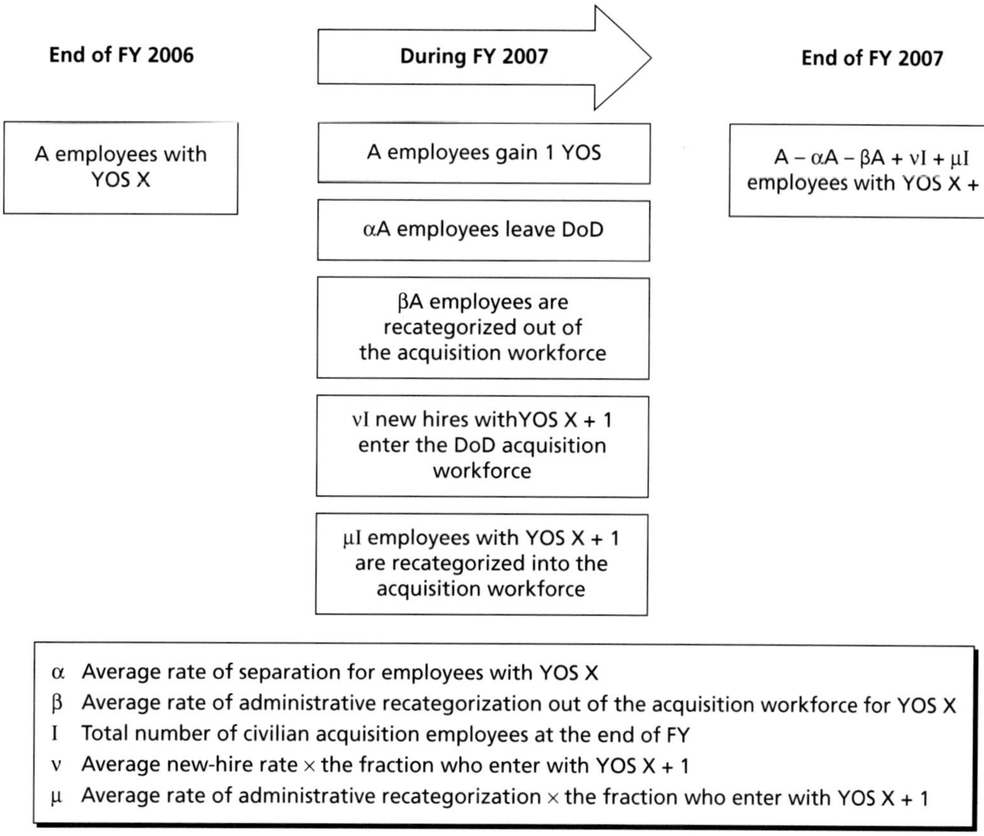

RAND *TR572-4.1*

[1] Alternatively, the model could focus on years relative to retirement rather than years of service. The same procedure would be followed, but all inventories and rate calculations would be based on an individual's years relative to retirement, as described in Chapter Three.

Acquisition Workforce Inventory Projection Model Overview

The starting point for the projection is the distribution of the AW in FY 2006 by YOS. To obtain projections for FY 2007, we use the following procedure. First, we let each YOS cell "age" by one year. In other words, individuals with YOS 1 in FY 2006 are moved into YOS 2 for FY 2007. We then account for the fact that some number of people might have left the DoD workforce or stayed in DoD but left the AW during this time. To do this, we calculate an expected continuation rate for each YOS. In our model, the expected continuation rate is simply 1 minus the average rate of separation from DoD over the past five years minus the average substantive recategorization rate over the past five years for that YOS. We calculate rates by YOS to account for the fact that separation rates tend to be much higher for workers with very few YOS and with many YOS. Finally, we account for new entrants into the AW. We next apply the five-year average rate of new hires to the total prior-year inventory to calculate the expected number of new hires between FY 2006 and FY 2007. We then calculate the expected YOS distribution of those new hires using historical averages for the YOS distribution of new hires. Similarly, we calculate the expected number and YOS distribution of "substantive" switches into the AW between FY 2006 and FY 2007.

The next two sections of this chapter provide detailed information on the model and how to use it. The final section provides examples of projections generated by the model. Readers who are not interested in the details of the model may wish to skip to that section.

Acquisition Workforce Projection Model Details

In this section, we provide detailed information on the AW inventory projection model that can be used to project the characteristics and size of the AW (or subpopulations of this workforce) in the future based on the size of the current inventory and historical turnover information. The actual workforce projection model is available from the authors upon request as a Microsoft Excel workbook. This model is designed to estimate how longevity-related characteristics of a workforce are likely to change over a period of several years. The results can be compared with sets of future workforce requirements developed as part of a comprehensive workforce planning exercise. The model can also be used to estimate how expected workforce characteristics might be affected by alternative human resource programs or policies or to assess what human resource programs might need to be implemented in order to achieve overall workforce goals.

It should be noted that this is only one version of the model that has been used internally at RAND. In this version, we allow the size of the AW to vary based on a set of assumptions regarding rates of separation and new hiring. Alternative versions of the model allow the user to either fix or vary the overall size of the workforce (under the assumption that vacant billets will be filled) in order to explore how other parameters vary.

Key inputs to the model are *beginning inventories* by YOS as well as *separation*, *recategorization (switch in* and *switch out)* and *new hire rates*.

Basic Configuration of the Model

The basic workforce characteristic depicted in the model is *year of service*. The model accepts as user input the *beginning inventory* of a workforce, distributed from YOS 0 to YOS 50.[2] The model uses *continuation rates* to calculate the number of workers in each YOS who are expected to remain in the workforce for an additional year. These *continuation rates* take into account losses due to *separations* and losses through substantive (but not administrative) *recategorizations* (switches out) of the AW. The model uses an *overall gain rate*, a *gain distribution* by YOS, and the previous year's end strength to calculate the number of workers in each YOS who are expected to enter the workforce.[3] These *gain rates* and *gain distributions* are separated into gains due to *new hires* and gains due to *recategorizations* (switches in) to the workforce. In summary, the model starts with a workforce as it looks at the end of FY 2006 and depicts how it might look at the end of each successive fiscal year.[4]

Figure 4.2 illustrates the basic configuration of the model. Column A of the sheet indicates the YOS. Column B contains the beginning inventory as it looked at the end of FY 2006.[5] Columns C through L contain the projected workforce at the ends of fiscal years 2007 through 2016. The user can enter an overall new-hire rate and switch-in rate in Q25 and Q26. Currently, the model uses the average rates between 2002 and 2006 as default inputs, but users can change the values in those cells. Line 55 indicates the expected end strength taking into consideration these rates, the continuation rates, and the new-hire and switch-in distributions by YOS. Figure 4.3 shows the bottom half of the sheet depicted in Figure 4.2.

Rows 55–67 provide summary statistics regarding the workforce, including distribution by YOS blocks and average YOS. Note, for example, that the average YOS drops from 18.8 years in FY 2006 to 17.9 years by FY 2016. This expected rejuvenation of the workforce can also be seen in the rows that depict the percentage of the workforce in various YOS groupings. The less-experienced percentages in YOS 1–3 and YOS 4–10 increase, while the more-experienced percentage in YOS 21–30 declines significantly over this period. Additionally, the last three lines provide an estimate of how much of the current (end of FY 2006) workforce will have separated or switched out by the end of each projected fiscal year. These data suggest that over half (57 percent) of the current workforce will have departed by FY 2016.

The spreadsheet depicted in Figures 4.2 and 4.3 contains formulas that calculate the expected workforces in each successive fiscal year. Each year group progresses through the matrix on a diagonal path. For example, the beginning inventory has 2,105 workers in YOS 15. About 96 percent of these workers continue on to YOS 16 in FY 2007. They are joined in

---

[2] YOS 50 contains all workers with more than 50 years of service. See "technical notes" for more information about YOS bins.

[3] An alternative version of this model requires the user to input end strengths and calculate the number of new hires necessary to achieve this desired end strength. This model differs by requiring the user to input new-hire and switch-in rates and calculate the end strengths that result.

[4] To be precise, these are the new hires or switches in who last until the end of the fiscal year in which they were hired/switched in. We do not observe in the data individuals who were hired in the course of a FY and quit before the end of that FY. If such short-term turnover were prevalent, then the number of people that the DoD would need to hire in order to achieve the number of "new hires" as we have defined it here would be higher.

[5] The data used in this example reflect the beginning inventory embedded in the workbook when it is initially supplied to users. This beginning inventory depicts the DoD-wide AW according to the acquisition flags included in the DoD Civilian Inventory File provided by DMDC.

Figure 4.2
Basic Configuration of the Model

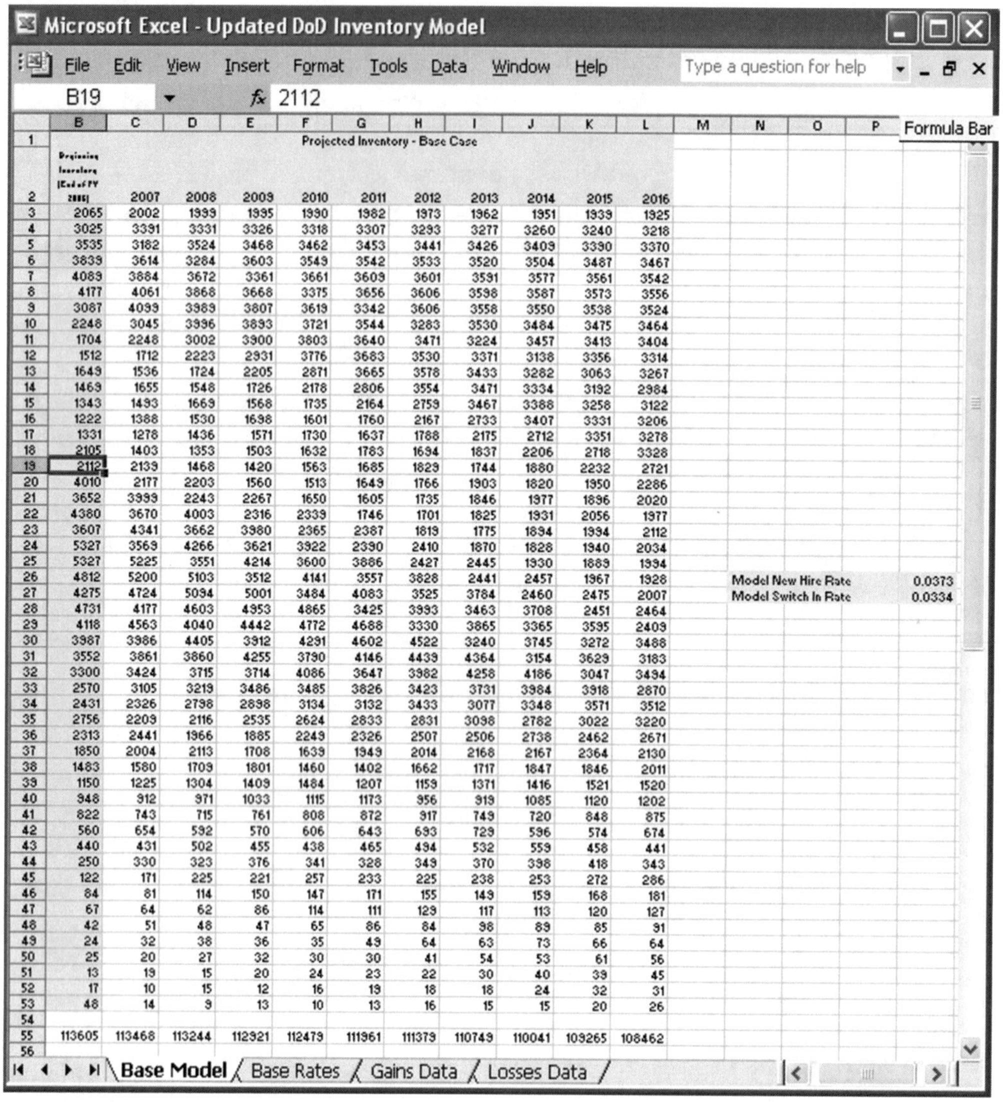

	B	C	D	E	F	G	H	I	J	K	L
1					Projected Inventory - Base Case						
2	[illegible]	2007	2008	2009	2010	2011	2012	2013	2014	2015	2016
3	2065	2002	1999	1995	1990	1982	1973	1962	1951	1939	1925
4	3025	3391	3331	3326	3318	3307	3293	3277	3260	3240	3218
5	3535	3182	3524	3468	3462	3453	3441	3426	3409	3390	3370
6	3839	3614	3284	3603	3549	3542	3533	3520	3504	3487	3467
7	4089	3884	3672	3361	3661	3609	3601	3591	3577	3561	3542
8	4177	4061	3868	3668	3375	3656	3606	3598	3587	3573	3556
9	3087	4099	3989	3807	3619	3342	3606	3558	3550	3538	3524
10	2248	3045	3996	3893	3721	3544	3283	3530	3484	3475	3464
11	1704	2248	3002	3900	3803	3640	3471	3224	3457	3413	3404
12	1512	1712	2223	2931	3776	3683	3530	3371	3138	3356	3314
13	1649	1536	1724	2205	2871	3665	3578	3433	3282	3063	3267
14	1469	1655	1548	1726	2178	2806	3554	3471	3334	3192	2984
15	1343	1493	1669	1568	1735	2164	2759	3467	3388	3258	3122
16	1222	1388	1530	1698	1601	1760	2167	2733	3407	3331	3206
17	1331	1278	1436	1571	1730	1637	1788	2175	2712	3351	3278
18	2105	1403	1353	1503	1632	1783	1694	1837	2206	2718	3328
19	2112	2139	1468	1420	1563	1685	1829	1744	1880	2232	2721
20	4010	2177	2203	1560	1513	1649	1766	1903	1820	1950	2286
21	3652	3999	2243	2267	1650	1605	1735	1846	1977	1896	2020
22	4380	3670	4003	2316	2339	1746	1701	1825	1931	2056	1977
23	3607	4341	3662	3980	2365	2387	1819	1775	1894	1994	2112
24	5327	3569	4266	3621	3922	2390	2410	1870	1828	1940	2034
25	5327	5225	3551	4214	3600	3886	2427	2445	1930	1889	1994
26	4812	5200	5103	3512	4141	3557	3828	2441	2457	1967	1928
27	4275	4724	5094	5001	3484	4083	3525	3784	2460	2475	2007
28	4731	4177	4603	4953	4865	3425	3993	3463	3708	2451	2464
29	4118	4563	4040	4442	4772	4688	3330	3865	3365	3595	2409
30	3987	3986	4405	3912	4291	4602	4522	3240	3745	3272	3488
31	3552	3861	3860	4255	3790	4146	4439	4364	3154	3629	3183
32	3300	3424	3715	3714	4086	3647	3982	4258	4186	3047	3494
33	2570	3105	3219	3486	3485	3826	3423	3731	3984	3918	2870
34	2431	2326	2798	2898	3134	3132	3433	3077	3348	3571	3512
35	2756	2209	2116	2535	2624	2833	2831	3098	2782	3022	3220
36	2313	2441	1966	1885	2249	2326	2507	2506	2738	2462	2671
37	1850	2004	2113	1708	1639	1949	2014	2168	2167	2364	2130
38	1483	1580	1709	1801	1460	1402	1662	1717	1847	1846	2011
39	1150	1225	1304	1409	1484	1207	1159	1371	1416	1521	1520
40	948	912	971	1033	1115	1173	956	919	1085	1120	1202
41	822	743	715	761	808	872	917	749	720	848	875
42	560	654	592	570	606	643	693	729	596	574	674
43	440	431	502	455	438	465	494	532	559	458	441
44	250	330	323	376	341	328	349	370	398	418	343
45	122	171	225	221	257	233	225	238	253	272	286
46	84	81	114	150	147	171	155	149	159	168	181
47	67	64	62	86	114	111	129	117	113	120	127
48	42	51	48	47	65	86	84	98	89	85	91
49	24	32	38	36	35	49	64	63	73	66	64
50	25	20	27	32	30	30	41	54	53	61	56
51	13	19	15	20	24	23	22	30	40	39	45
52	17	10	15	12	16	19	18	18	24	32	31
53	48	14	9	13	10	13	16	15	15	20	26
54											
55	113605	113468	113244	112921	112479	111961	111379	110749	110041	109265	108462
56											

RAND *TR572-4.2*

YOS 16 by about 0.3 percent of the 4,233 new hires in FY 2007 and 3 percent of the 3,798 switches in during FY 2007 (those who enter with 15 years of previous service), resulting in a YOS 16 population of 2,139 by the end of FY 2007. These 2,139 workers are similarly "aged" to arrive at the YOS 17 population of 2,203 by the end of FY 2008. (See the shaded diagonal in the close-up view of the spreadsheet in Figure 4.4).

The formulas contained in the workforce sheet use continuation rates and the distribution of gains (new hires and switches in) by year of service. These rates are contained in a separate sheet in the workbook. Note, in Figures 4.2 through 4.4, that sheets in an Excel workbook are identified by a row of tabs along the bottom of the workbook window. The material depicted in Figures 4.2 through 4.4 is in the sheet labeled "Base Model." The accompanying rates are in the sheet labeled "Base Rates."

Figure 4.3
Summary Statistics

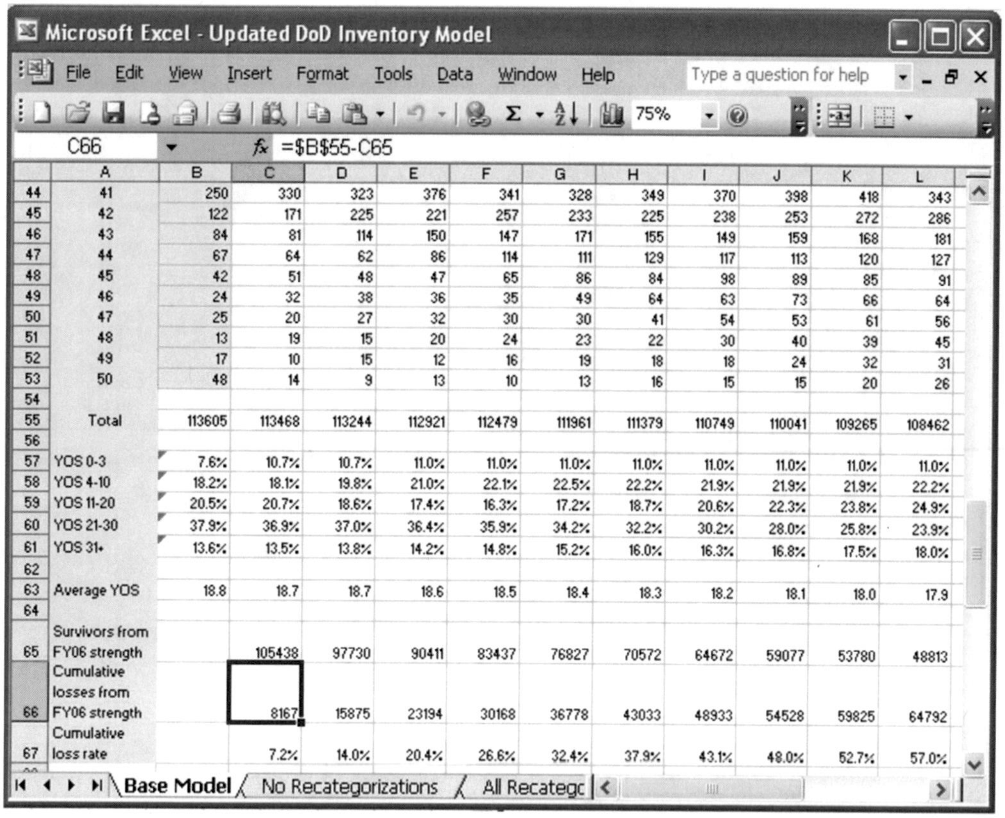

	A	B	C	D	E	F	G	H	I	J	K	L
44	41	250	330	323	376	341	328	349	370	398	418	343
45	42	122	171	225	221	257	233	225	238	253	272	286
46	43	84	81	114	150	147	171	155	149	159	168	181
47	44	67	64	62	86	114	111	129	117	113	120	127
48	45	42	51	48	47	65	86	84	98	89	85	91
49	46	24	32	38	36	35	49	64	63	73	66	64
50	47	25	20	27	32	30	30	41	54	53	61	56
51	48	13	19	15	20	24	23	22	30	40	39	45
52	49	17	10	15	12	16	19	18	18	24	32	31
53	50	48	14	9	13	10	13	16	15	15	20	26
54												
55	Total	113605	113468	113244	112921	112479	111961	111379	110749	110041	109265	108462
56												
57	YOS 0-3	7.6%	10.7%	10.7%	11.0%	11.0%	11.0%	11.0%	11.0%	11.0%	11.0%	11.0%
58	YOS 4-10	18.2%	18.1%	19.8%	21.0%	22.1%	22.5%	22.2%	21.9%	21.9%	21.9%	22.2%
59	YOS 11-20	20.5%	20.7%	18.6%	17.4%	16.3%	17.2%	18.7%	20.6%	22.3%	23.8%	24.9%
60	YOS 21-30	37.9%	36.9%	37.0%	36.4%	35.9%	34.2%	32.2%	30.2%	28.0%	25.8%	23.9%
61	YOS 31+	13.6%	13.5%	13.8%	14.2%	14.8%	15.2%	16.0%	16.3%	16.8%	17.5%	18.0%
62												
63	Average YOS	18.8	18.7	18.7	18.6	18.5	18.4	18.3	18.2	18.1	18.0	17.9
64												
65	Survivors from FY06 strength		105438	97730	90411	83437	76827	70572	64672	59077	53780	48813
66	Cumulative losses from FY06 strength		8167	15875	23194	30168	36778	43033	48933	54528	59825	64792
67	Cumulative loss rate		7.2%	14.0%	20.4%	26.6%	32.4%	37.9%	43.1%	48.0%	52.7%	57.0%

RAND *TR572-4.3*

Technical Notes

In this section, we provide additional information about elements of the Excel workbook.

YOS Bins

In an inventory matrix, the bins identify the population with *n* years of service completed. Thus, YOS 0 identifies the population with less than one year of service as of the end of a given fiscal year. YOS 1 identifies the population with at least one but less than two years of service.

In a set of continuation rates, the rate for YOS *n* indicates the rate of continuation for workers with *n* years of service. Thus, the continuation rate for YOS 10 in FY 2007 indicates the proportion of workers with YOS 10 in FY 2007 who continue into YOS 11 in FY 2008. People do not continue as part of the AW for two reasons: because they separate from DoD employment or because they switch out of the AW. The continuation rates included in this model are equal to 1 minus the average separation rate for workers with *n* YOS over five years minus the average switching-out rate over five years (2002 to 2006).[6]

[6] We include substantive but not administrative switches out of the AW.

Figure 4.4
Diagonal Progression of a Year Group Through the Model

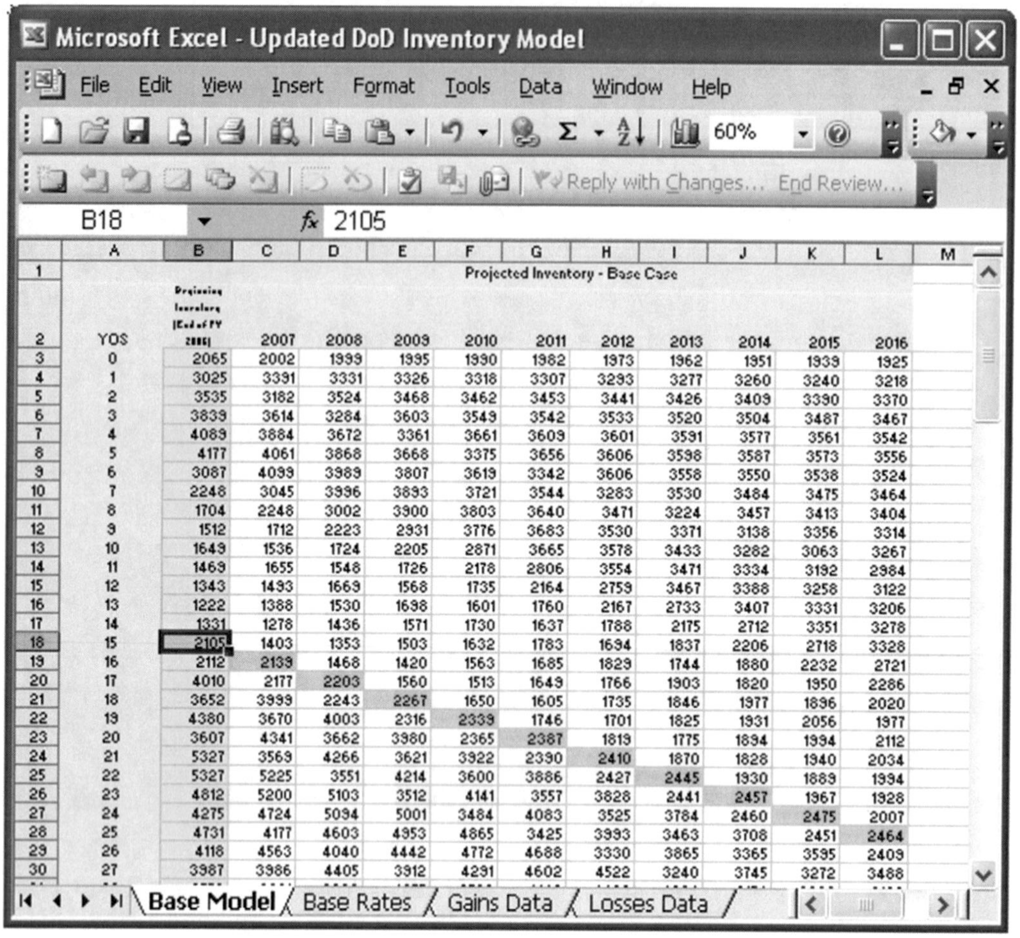

Projected Inventory - Base Case

YOS	Beginning Inventory (End of FY 2006)	2007	2008	2009	2010	2011	2012	2013	2014	2015	2016
0	2065	2002	1999	1995	1990	1982	1973	1962	1951	1939	1925
1	3025	3391	3331	3326	3318	3307	3293	3277	3260	3240	3218
2	3535	3182	3524	3468	3462	3453	3441	3426	3409	3390	3370
3	3839	3614	3284	3603	3549	3542	3533	3520	3504	3487	3467
4	4089	3884	3672	3361	3661	3609	3601	3591	3577	3561	3542
5	4177	4061	3868	3668	3375	3656	3606	3598	3587	3573	3556
6	3087	4099	3989	3807	3619	3342	3606	3558	3550	3538	3524
7	2248	3045	3996	3893	3721	3544	3283	3530	3484	3475	3464
8	1704	2248	3002	3900	3803	3640	3471	3224	3457	3413	3404
9	1512	1712	2223	2931	3776	3683	3530	3371	3138	3356	3314
10	1649	1536	1724	2205	2871	3665	3578	3433	3282	3063	3267
11	1469	1655	1548	1726	2178	2806	3554	3471	3334	3192	2984
12	1343	1493	1669	1568	1735	2164	2759	3467	3388	3258	3122
13	1222	1388	1530	1698	1601	1760	2167	2733	3407	3331	3206
14	1331	1278	1436	1571	1730	1637	1788	2175	2712	3351	3278
15	2105	1403	1353	1503	1632	1783	1694	1837	2206	2718	3328
16	2112	2139	1468	1420	1563	1685	1829	1744	1880	2232	2721
17	4010	2177	2203	1560	1513	1649	1766	1903	1820	1950	2286
18	3652	3999	2243	2267	1650	1605	1735	1846	1977	1896	2020
19	4380	3670	4003	2316	2339	1746	1701	1825	1931	2056	1977
20	3607	4341	3662	3980	2365	2387	1819	1775	1894	1994	2112
21	5327	3569	4266	3621	3922	2390	2410	1870	1828	1940	2034
22	5327	5225	3551	4214	3600	3886	2427	2445	1930	1889	1994
23	4812	5200	5103	3512	4141	3557	3828	2441	2457	1967	1928
24	4275	4724	5094	5001	3484	4083	3525	3784	2460	2475	2007
25	4731	4177	4603	4953	4865	3425	3993	3463	3708	2451	2464
26	4118	4563	4040	4442	4772	4688	3330	3865	3365	3595	2409
27	3987	3986	4405	3912	4291	4602	4522	3240	3745	3272	3488

RAND *TR572-4.4*

In a set of gain distribution rates, the rate for YOS *n* indicates the average proportion of annual gains in the specified YOS. Thus, the new-hire distribution rate for YOS 5 in FY 2008 indicates the proportion of all FY 2008 new hires who will have more than four but less than or equal to five years of service at the end of FY 2008. The switch-in distribution rates function in the same manner.

Survivors from FY 2006 Strength

The calculations used here are made by compounding the continuation rates along a diagonal in the continuation rate matrix, multiplying the compound, multiyear continuation rate by the appropriate YOS bin in the FY 2006 beginning inventory and summing the products.

Loss and Continuation Rates

An annual loss rate, by YOS, is computed by dividing total losses during a year in each YOS by the number of workers in the YOS at the beginning of the year. A continuation rate is the reciprocal of a loss rate.

Users of the model can substitute organization-specific or occupation-specific loss rates in place of the DoD-wide rates installed in the model by pasting the new rates into cells B3 through B52 (corresponding to YOS 0 through 49) and C3 through C52 in each applicable rates sheet. When these rate substitutions are made, the model automatically computes the continuation rates in cells D3 through D53 of the Base Rates sheet (the FY 2007 column). Additionally, to truncate the model at YOS 50, the continuation rate for YOS 50 should remain 0.

Gain Distributions

Gain distribution rates are calculated from the counts of annual gains by YOS appearing in the Gains Data spreadsheet. Gain distribution rates are calculated separately for new hires into DoD and switches into the AW. The model converts these counts into rates and the current model averages the rates over five years as well. Users can substitute organization-specific or occupation-specific gain counts in place of the DoD-wide rates installed in the model by pasting the new counts into the Gains Data sheet, then copying the values from this sheet into B66 through B116 and O66 through O116 in the Base Rates sheet.

Gain Rates

An annual gain rate is computed by dividing total gains during a year by the total number of workers at the beginning of the year. These rates are computed separately for new hires and switches in. The model assumes that these gain rates remain constant over time. The current model uses average gain rates over five years to estimate these gain rates, but users are free to substitute any gain rates to determine how the model is affected (e.g., to see how increasing the new-hire rate to 5 percent would change workforce projections, simply enter 5 in the cell specifying *Model New Hire Rate*).

Manipulating the Inventory Projection Model

The workbook contains a beginning inventory for the DoD-wide AW equal to the end strength for FY 2006. The continuation rates, model gain rates, and gain distribution rates are based on averages of DoD-wide AW civil service gains and losses between 2002 and 2006. With some minor modifications, the model can also be used to examine subsets of the AW. In addition, the model can be modified to explore alternative assumptions about future workforce turnover or workforce management practices. Customization can be done by modifying the beginning inventory and modifying the rates.

Modifying the Beginning Inventory

To analyze a different population, the model substitutes a beginning inventory drawn from data on a workforce of interest. The workforce of interest might be the entire workforce in the user's organization or specific subgroups within the workforce, such as the AW, a specific occupation, or perhaps the population with advanced degrees in a specific occupation.

Changing Gain Rates

Two gain rates are used in the model: the new-hire rate and the rate of administrative recategorization into the AW—i.e., the switch-in rate. The new-hire rate feeding the model is entered in cell Q26 in the Base Model tab. The switch-in rate is entered in cell Q27. The spreadsheet is set up to automatically feed the average rates over the past five years into those cells. However, the user may choose to enter a different rate. For example, one might choose to focus on the actual rates in the most recent year, or play with the rates to identify the rate that would be needed to maintain the current workforce.

Changing the Distribution of Gains

This type of change would simulate a shift in recruiting emphasis. For example, a strategy of placing greater emphasis on hiring experienced workers from other agencies or individuals with prior military experience would result in a smaller proportion of gains in YOS 1 and greater proportion in YOS 2 and above.

The distribution of gains being applied in the model is found in Column V of the Gains Data sheet based on the average over the last five fiscal years. To change the distribution of gains, a user would open the Gains Data sheet and scroll down to the gain distribution rates on the lower half of the sheet. Select the YOS cells likely to be affected by the program or policy alternative being simulated. Substitute new rates for the ones supplied in the model. Bear in mind that whenever one cell is changed, offsetting changes must be made so that the distribution rates sum to one.

Changing Continuation Rates

Continuation rates, found in columns D through M on the Base Rates sheet, are a key driver of the projection. There is a separate continuation rate for each YOS bin and for each fiscal year. The model assumes that for each YOS bin, the continuation rate will be 1 minus the average historical separation rate for that YOS bin minus the average historical switch-out rate for that YOS bin. However, the model is set up to allow users to change those separation rates.

This type of change would simulate the effects of increasing or decreasing retention of selected parts of the workforce. For example, paying retention bonuses to workers in selected years of service would likely increase continuation rates in those years. Likewise, paying Voluntary Separation Incentive Pay (VSIP) to the workforce would likely decrease continuation rates, especially for YOS groups in the retirement-eligible range.

Estimating Policy Effects

The model does not have an ability to estimate how much the continuation or gain distribution rates might change as a result of a policy or program change. The user must estimate the direction and magnitude of the effect. One basis for such estimates, if available, would be rates derived from some previous period in which the same or similar policies were in force.[7]

---

[7] In some very sophisticated workforce modeling applications, such as those used by some services for military force programming, retention behaviors have been estimated as a function of the alternative income streams faced by those leaving service and those remaining in service. In such applications, the retention effects of any policy that can be monetarized are readily simulated. However, developing such underlying behavioral models is beyond the scope of work supporting the relatively simple inventory projection model provided here.

Acquisition Workforce Inventory Projections Using the Model

In this section, we apply the inventory projection model to civilian acquisition personnel inventory data from the end of FY 2006 and use average rates of separation over the prior five fiscal years to generate the ten-year inventory projection for the AW shown in Figure 4.5.[8] The projection suggests that if historical gain and separation rates by YOS hold over the next decade, the civilian AW will shrink slightly by about 5,000 over the next ten years. The model can be used by managers to explore options for changing the outcome as well. For example, managers can see that increasing the new-hire rate by as little as 0.5 percent (to 4.2 percent) could allow DoD to maintain the AW at its current size.

As discussed in Chapter Three, a larger number of DoD employees are recategorized into and out of the AW in any given year and the number of recategorizations into the AW typically exceed the number of recategorizations out of the workforce. We also concluded that well over half of all recategorizations appear to be administrative in nature. It is therefore logical to expect that the way in which recategorizations are treated will have important implications for workforce projections.

The projection in Figure 4.5 includes substantive but not administrative recategorizations. In forecasting separations from and movement into the AW, the projection assumes that there will be some gains and losses to/from other parts of DoD and that those gains and losses can be best projected by looking at historical rates of substantive (but not administrative) recategorizations into and out of the AW.[9]

Figure 4.6 illustrates how sensitive future workforce projections are to the assumptions made about recategorizations.

As illustrated in Figure 4.6, a projection that accounts for past administrative recategorizations and assumes those will continue in the future results in a projected AW of more than 220,000 by FY 2016. In contrast, a projection that assumes there will be no recategorizations in the future (i.e., that all gains and losses will result from new hires or losses to DoD) results in a projected workforce of 82,724 by FY 2016—substantially smaller than that generated by our base model that includes substantive recategorizations.

Readers may also be interested in how sensitive the model is to changes in assumptions about the rate of new hiring. Indeed, some DoD managers with whom this model was discussed mentioned that they would prefer to use a different assumption about hiring rates. The model is well equipped to handle such adjustments, as described earlier in this chapter. The base model assumes that the future rate of hiring will equal the five-year historical average of 3.7 percent. DoD-wide, this rate has varied from a low of 3.4 percent in FY 2006 to a

---

[8] As noted above, users may wish to use different separation rates in the model—for example, the three-year average separation rate or simply the separation rate from the prior year. The model can be easily adjusted to use these average separation rates. There are pros and cons of different assumptions, and managers with an intimate knowledge of the historical trends experienced by their particular workforce will be in the best position to judge which historical rates are the most appropriate inputs into the model. We do caution, however, that because the separation rates are calculated for each YOS group, the rates for a single year may be highly variable for small subsegments of the workforce (e.g., specific career fields or occupational groups within specific commands).

[9] This assumption would imply either that there are no administrative recategorizations in the future or that the projections do not account for future recategorizations.

Figure 4.5
Projection of the Size of the DoD Civilian Acquisition Workforce, FY 2006–2016

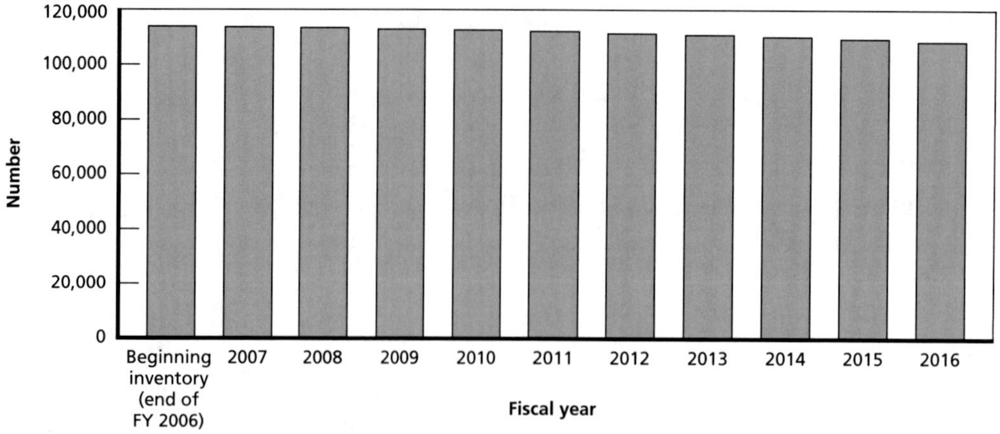

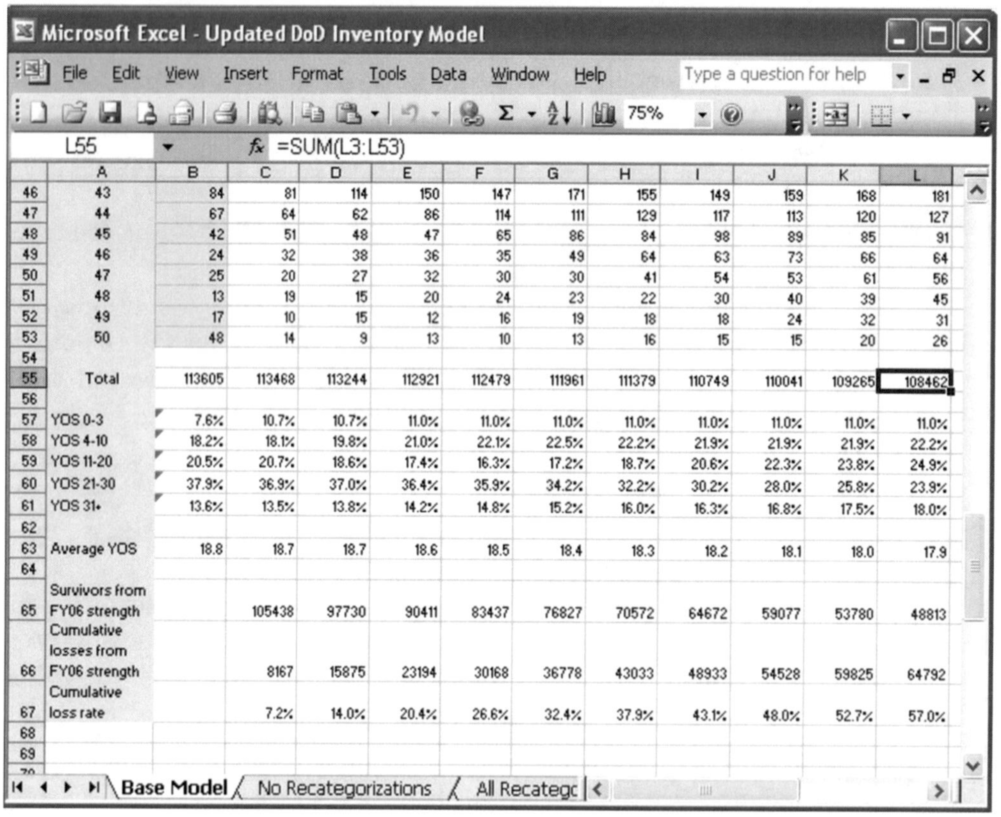
Microsoft Excel - Updated DoD Inventory Model

L55 =SUM(L3:L53)

	A	B	C	D	E	F	G	H	I	J	K	L
46	43	84	81	114	150	147	171	155	149	159	168	181
47	44	67	64	62	86	114	111	129	117	113	120	127
48	45	42	51	48	47	65	86	84	98	89	85	91
49	46	24	32	38	36	35	49	64	63	73	66	64
50	47	25	20	27	32	30	30	41	54	53	61	56
51	48	13	19	15	20	24	23	22	30	40	39	45
52	49	17	10	15	12	16	19	18	18	24	32	31
53	50	48	14	9	13	10	13	16	15	15	20	26
54												
55	Total	113605	113468	113244	112921	112479	111961	111379	110749	110041	109265	108462
56												
57	YOS 0-3	7.6%	10.7%	10.7%	11.0%	11.0%	11.0%	11.0%	11.0%	11.0%	11.0%	11.0%
58	YOS 4-10	18.2%	18.1%	19.8%	21.0%	22.1%	22.5%	22.2%	21.9%	21.9%	21.9%	22.2%
59	YOS 11-20	20.5%	20.7%	18.6%	17.4%	16.3%	17.2%	18.7%	20.6%	22.3%	23.8%	24.9%
60	YOS 21-30	37.9%	36.9%	37.0%	36.4%	35.9%	34.2%	32.2%	30.2%	28.0%	25.8%	23.9%
61	YOS 31+	13.6%	13.5%	13.8%	14.2%	14.8%	15.2%	16.0%	16.3%	16.8%	17.5%	18.0%
62												
63	Average YOS	18.8	18.7	18.7	18.6	18.5	18.4	18.3	18.2	18.1	18.0	17.9
64												
65	Survivors from FY06 strength		105438	97730	90411	83437	76827	70572	64672	59077	53780	48813
66	Cumulative losses from FY06 strength		8167	15875	23194	30168	36778	43033	48933	54528	59825	64792
67	Cumulative loss rate		7.2%	14.0%	20.4%	26.6%	32.4%	37.9%	43.1%	48.0%	52.7%	57.0%
68												
69												

Base Model / No Recategorizations / All Recatego

RAND *TR572-4.5*

high of 4.1 percent in FY 2002. At the service level, the variation in hiring rates is even greater. In Figure 4.7, we demonstrate how the projection changes when the assumptions about the rates of new hires change. If we assume that the new-hire rate in the future will be one-half a percentage point higher than the five-year historical average (or 4.2 percent), as shown in the first spreadsheet, the projected inventory in FY 2016 is 114,087 rather than 108,462. Similarly,

Figure 4.6
Projection of the Size of the DoD Civilian Acquisition Workforce,Under Different Assumptions About Recategorization, FY 2006–2016

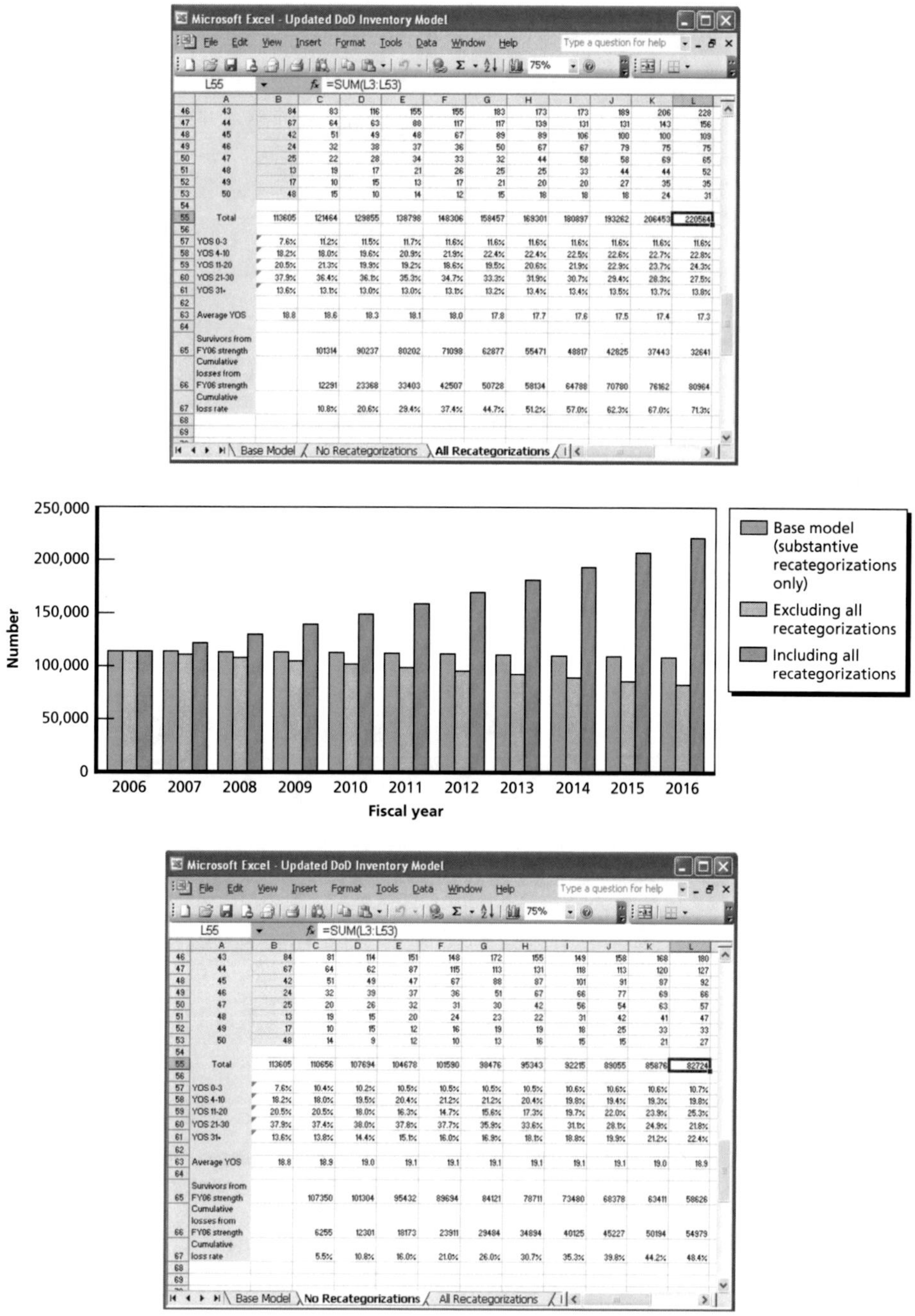

RAND *TR572-4.6*

if we assume that the future hiring rate is one-half a percentage point lower (or 3.2 percent), as shown on the second spreadsheet, then the FY 2016 inventory is 102,506.

The inventory projection is a simple yet powerful tool for workforce managers and strategic planners. Although we provide the projection here for the total DoD civilian AW, it is

Figure 4.7
Inventory Projections Under Different Hiring Rate Assumptions

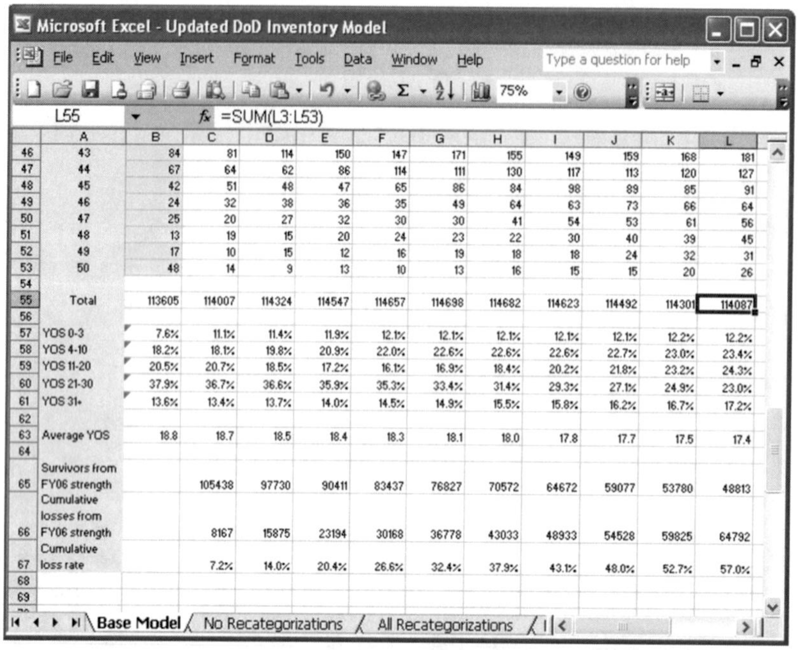

	A	B	C	D	E	F	G	H	I	J	K	L
46	43	84	81	114	150	147	171	155	149	159	168	181
47	44	67	64	62	86	114	111	130	117	113	120	127
48	45	42	51	48	47	65	86	84	98	89	85	91
49	46	24	32	38	36	35	49	64	63	73	66	64
50	47	25	20	27	32	30	30	41	54	53	61	56
51	48	13	19	15	20	24	23	22	30	40	39	45
52	49	17	10	15	12	16	19	18	18	24	32	31
53	50	48	14	9	13	10	13	16	15	15	20	26
54												
55	Total	113605	114007	114324	114547	114657	114698	114682	114623	114492	114301	114087
56												
57	YOS 0-3	7.6%	11.1%	11.4%	11.9%	12.1%	12.1%	12.1%	12.1%	12.1%	12.2%	12.2%
58	YOS 4-10	18.2%	18.1%	19.8%	20.9%	22.0%	22.6%	22.6%	22.6%	22.7%	23.0%	23.4%
59	YOS 11-20	20.5%	20.7%	18.5%	17.2%	16.1%	16.9%	18.4%	20.2%	21.8%	23.2%	24.3%
60	YOS 21-30	37.9%	36.7%	36.6%	35.9%	35.3%	33.4%	31.4%	29.3%	27.1%	24.9%	23.0%
61	YOS 31+	13.6%	13.4%	13.7%	14.0%	14.5%	14.9%	15.5%	15.8%	16.2%	16.7%	17.2%
62												
63	Average YOS	18.8	18.7	18.5	18.4	18.3	18.1	18.0	17.8	17.7	17.5	17.4
64												
65	Survivors from FY06 strength		105438	97730	90411	83437	76827	70572	64672	59077	53780	48813
66	Cumulative losses from FY06 strength		8167	15875	23194	30168	36778	43033	48933	54528	59825	64792
67	Cumulative loss rate		7.2%	14.0%	20.4%	26.6%	32.4%	37.9%	43.1%	48.0%	52.7%	57.0%
68												
69												

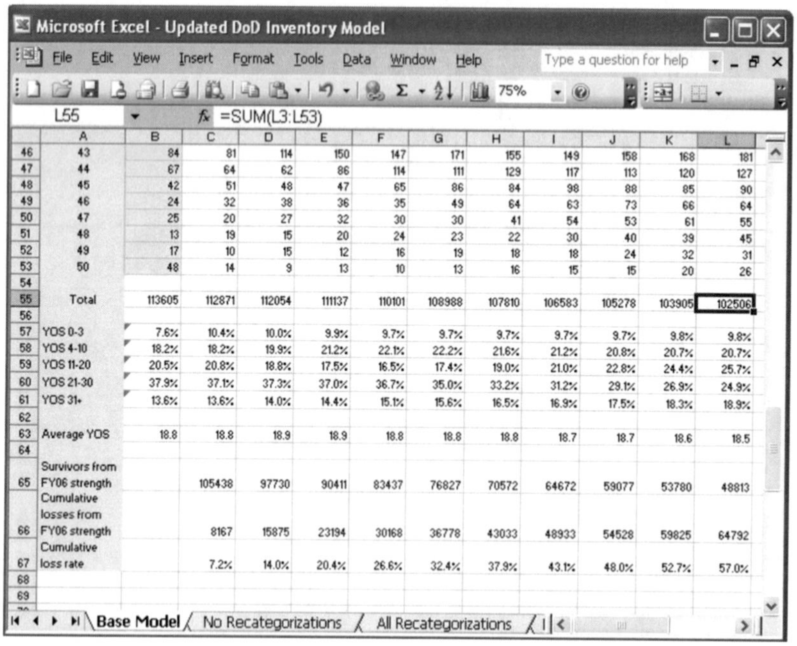

	A	B	C	D	E	F	G	H	I	J	K	L
46	43	84	81	114	150	147	171	155	149	158	168	181
47	44	67	64	62	86	114	111	129	117	113	120	127
48	45	42	51	48	47	65	86	84	98	88	85	90
49	46	24	32	38	36	35	49	64	63	73	66	64
50	47	25	20	27	32	30	30	41	54	53	61	55
51	48	13	19	15	20	24	23	22	30	40	39	45
52	49	17	10	15	12	16	19	18	18	24	32	31
53	50	48	14	9	13	10	13	16	15	15	20	26
54												
55	Total	113605	112871	112054	111137	110101	108988	107810	106583	105278	103905	102506
56												
57	YOS 0-3	7.6%	10.4%	10.0%	9.9%	9.7%	9.7%	9.7%	9.7%	9.7%	9.8%	9.8%
58	YOS 4-10	18.2%	18.2%	19.9%	21.2%	22.1%	22.2%	21.6%	21.2%	20.8%	20.7%	20.7%
59	YOS 11-20	20.5%	20.8%	18.8%	17.5%	16.5%	17.4%	19.0%	21.0%	22.8%	24.4%	25.7%
60	YOS 21-30	37.9%	37.1%	37.3%	37.0%	36.7%	35.0%	33.2%	31.2%	29.1%	26.9%	24.9%
61	YOS 31+	13.6%	13.6%	14.0%	14.4%	15.1%	15.6%	16.5%	16.9%	17.5%	18.3%	18.9%
62												
63	Average YOS	18.8	18.8	18.9	18.9	18.8	18.8	18.8	18.7	18.7	18.6	18.5
64												
65	Survivors from FY06 strength		105438	97730	90411	83437	76827	70572	64672	59077	53780	48813
66	Cumulative losses from FY06 strength		8167	15875	23194	30168	36778	43033	48933	54528	59825	64792
67	Cumulative loss rate		7.2%	14.0%	20.4%	26.6%	32.4%	37.9%	43.1%	48.0%	52.7%	57.0%
68												
69												

RAND *TR572-4.7*

possible to make similar projections for segments of the workforce. For example, it is possible to perform the analysis by service or by acquisition career field. When making projections for smaller segments of the workforce, managers must carefully consider how they would like to account for movement between segments, if at all. For example, should a service-level projection account for individuals who switch services?

It is worth noting that in our inventory projection model for the AW, we account for substantive, but not administrative, switches into and out of the AW. Because the rate of "recategorization in" (or internal hires) has been higher than the rate of "recategorization out" (or internal losses), accounting for these switches provides a more optimistic (and we would argue, more accurate) picture of the future workforce size.

In our application of the projection model, we have made some specific assumptions with which some might disagree. However, the model is flexible and would allow managers to explore the implications of alternative assumptions. For example, we have assumed that the new-hire rate over the next ten years would equal the average new-hire rate over the past five years (3.7 percent). This five-year average factors in a relatively high rate of new hires (4.1 percent) in FY 2002. Managers may prefer to perform the projection assuming a lower rate of new hires, such as that observed in FY 2006 (3.4 percent). Alternatively, managers may configure the model to determine the rate of new hires required to maintain a fixed or specified set of AW sizes. As described in this chapter, it is relatively easy for managers to modify the projections in this way.

In the next chapter, we focus specifically on AW senior executive service members.

CHAPTER FIVE

The Careers of Acquisition Workforce Senior Executive Service Employees

This chapter focuses on the careers of a very specific subset of the AW—the 454 Senior Executive Service employees in the AW as of the September 30, 2006, inventory. We were interested in learning about these individuals and their career developments to gain insight into how AW leadership has developed and evolved. In this chapter we present a descriptive overview of the DoD SES acquisition workforce, their career trajectories, and interservice mobility. We also examine the characteristics of individuals who enter the civilian AW at high grade levels.

Descriptive Overview of the DoD SES Acquisition Workforce

We begin by providing some basic descriptive information on this population. Of the 454 people, 380 are males. They ranged in age from 34 to 78, but more than half were in their 50s. Of these, 163 worked for the Navy, 157 worked for the Army, 54 worked for the Air Force, 78 worked for the OSD and defense agencies, and two worked for the Marine Corps. Not surprisingly, as shown in Figure 5.1, more than half of the AW SESs worked in the Washington, D.C., area.

Figure 5.1 was derived using the Metropolitan Statistical Area code in the individuals' civilian inventory record.[1]

Career Trajectories of Current SES Members

Using each individual's scrambled SSN, we then traced these individuals' careers back to September 30, 1992, by means of the annual inventory files. Of the 454 people, 393 were DoD civilian employees on September 30, 1992. Figure 5.2 shows the grade levels of the 454 FY 2006 AW SESs for each year dating back to September 30, 1992.

Of the 454 people, 137 were GS- or GM-15s as of September 30, 1992. Sixty-one had already obtained SES status. One remarkable individual was a GS-5 as of September 30, 1992.

[1] Patuxent River, however, is not in a census bureau Metropolitan Statistical Area. Patuxent River workers were identified by their duty zip code, 20670.

Figure 5.1
The Work Location of Acquisition Workforce SESs, September 2006

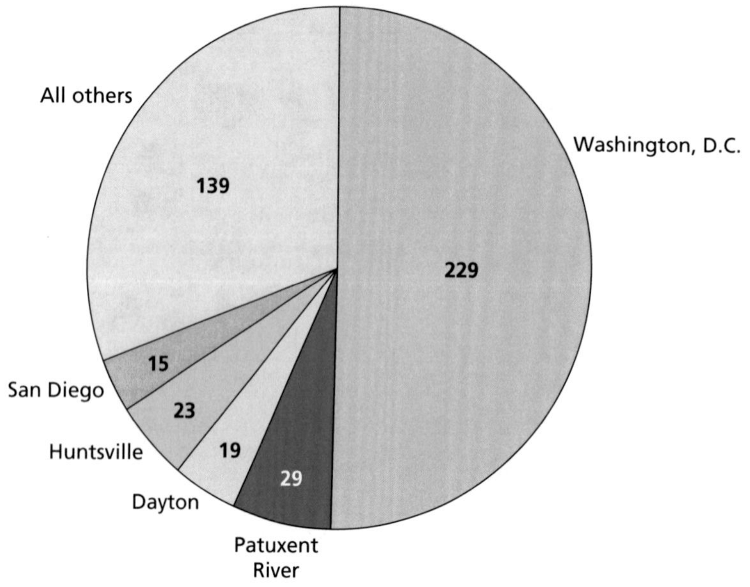

RAND *TR572-5.1*

Figure 5.2
Historical Grade Levels of FY 2006 Acquisition Workforce SESs

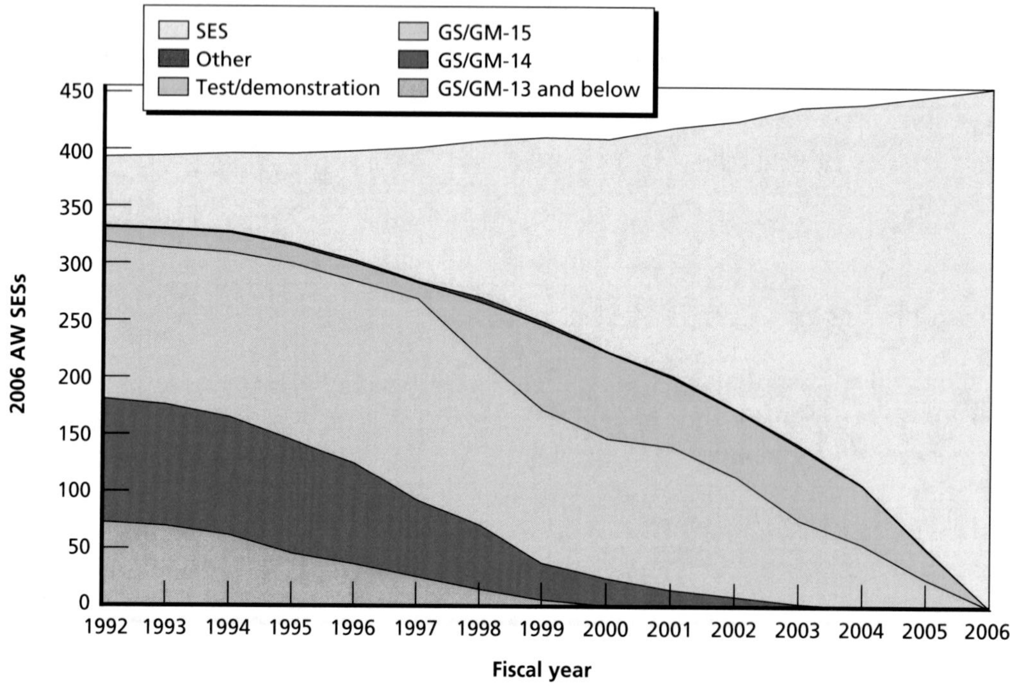

RAND *TR572-5.2*

Of the 454 FY 2006 AW SESs, 307 had at least one year between 1992 and 2005 during which they were a GS- or GM-15.[2]

For some SESs who were GS- or GM-15s, we observe the beginning of their tenure as a GS/GM-15 (as well as, in all likelihood, the end of that tenure—because we know that they were SESs on September 30, 2006). We refer to these cases as "fully observed" GS/GM-15 durations. These individuals, on average, were GS/GM-15s for 4.4 years, but the durations varied widely, as shown in Figure 5.3.

Another population was *truncated,* i.e., they were already GS/GM-15s in 1992, so we do not know when they first attained that status. These individuals were GS/GM-15s for 6.9 years in our data, on average, but we know this value is a lower bound on their actual mean duration in that grade.

The variability of future SESs' durations as GS/GM-15s is striking. Some individuals moved up the personnel schedule rapidly; others lingered for many years as a 15. There is much less variability in time-in-grade for military officers (Schirmer et al., 2006).

Mobility of Acquisition Workforce SES Members

Of the 454 FY 2006 AW SESs, 398 did not work in multiple military services (or bureaus) between 1992 and 2006. The AW appears to value and reward depth, not breadth, of experience for civilian employees.

Figure 5.3
Duration of Acquisition Workforce SESs' Time as GS/GM-15s

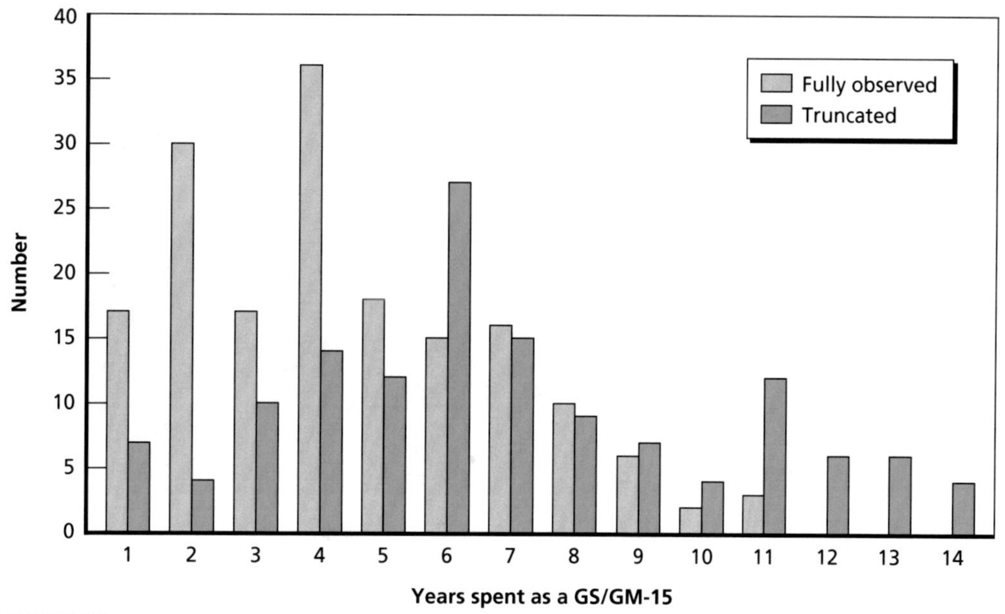

RAND *TR572-5.3*

[2] Sixty-one were SESs throughout. Others never were GS/GM-15s. For example, they were part of various test or demonstration pay programs before becoming SESs, but they were not hired in at the SES level.

To the extent that the FY 2006 AW SESs did change services, their moves tended to be from the Department of the Navy (Navy and Marine Corps) toward OSD and the Army, as shown in Figure 5.4.

We refer to an individual as having been "exported" by a military service if he or she moved out of that service between 1992 and 2006, irrespective of his or her grade at the time of the move. If, for example, an individual in the data set worked for the Army in 1992–1996, moved to the Air Force for 1997–2000, then moved to OSD for 2001–2006, becoming an OSD SES in 2003, this individual would be tallied as having been "exported" by both the Army (1997 move to the Air Force) and the Air Force (2001 move to OSD) and "imported" by both the Air Force (1997) and OSD (2001). (The sum of total imports equals the sum of total exports.)

Figure 5.4 does not display moves within the DoN. Six civilians moved from the Navy to the Marine Corps; four moved from the Marine Corps to the Navy.

Lateral Entry into Senior DoD Civilian Acquisition Workforce Positions

Although the norm has been for AW SESs to be long-time DoD civilian employees, there are some exceptions. Figure 5.5 presents information on 61 FY 2006 AW SESs whom we colloquially label "surprising" SESs. These individuals were hired into the civilian AW at high grades (GS-13 and higher) in the years indicated on the x-axis of the figure. Indeed, 35 of the 61 had their first DoD civilian position at the SES level.

Figure 5.4
Interservice Moves of FY 2006 Acquisition Workforce SESs, 1992–2006

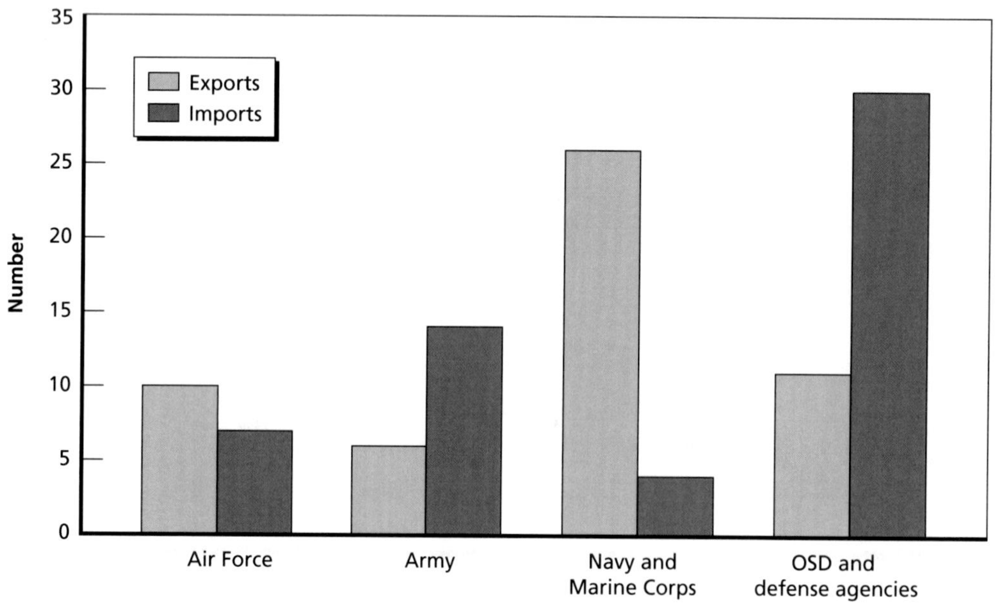

RAND *TR572-5.4*

Figure 5.5
The "Surprising" SESs

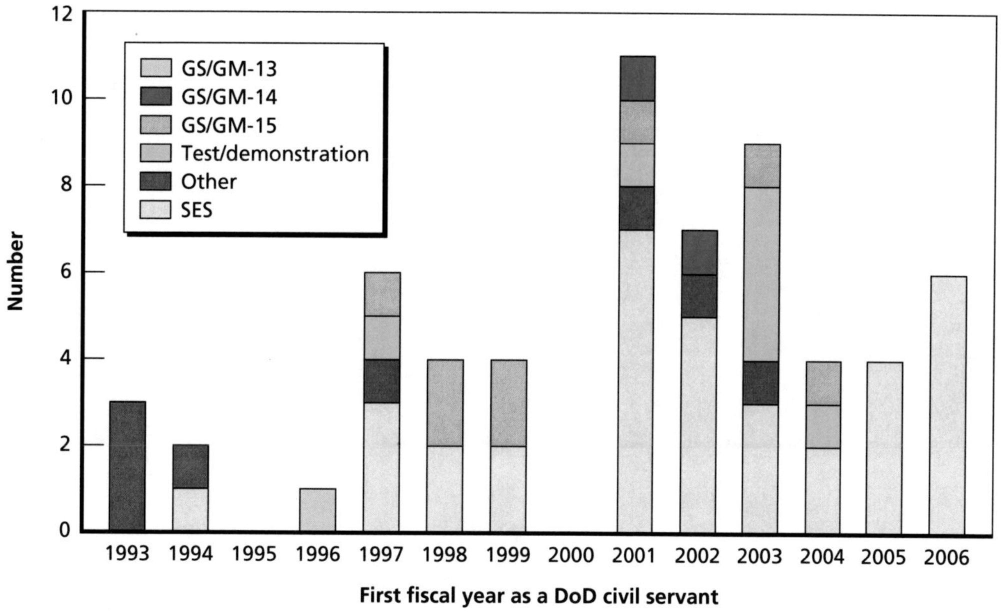

RAND *TR572-5.5*

One explanation for the surprising SESs is prior military service. Forty-eight of the 61 FY 2006 AW SESs shown in Figure 5.5 were previously military officers, according to the WEX. Figure 5.6 displays the military service of the now-civilian AW SESs in question. Not surpris-

Figure 5.6
The Military Backgrounds of the Surprising SESs

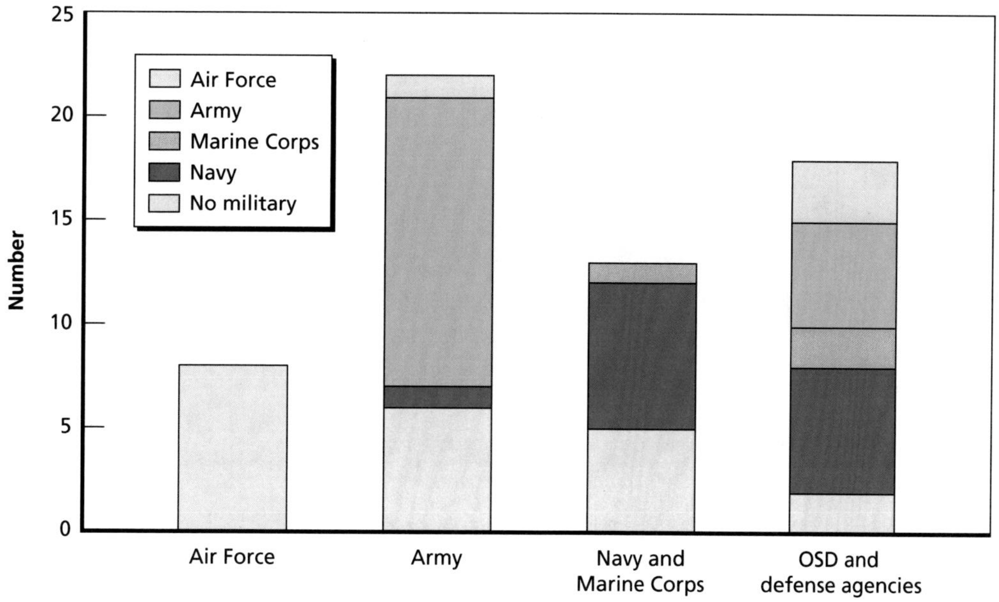

RAND *TR572-5.6*

ingly, military officers tended to become civilian employees of the specific military service in which they had served (with OSD and the defense agencies hiring officers from all military services).

This discussion of the prior military service of FY 2006 AW SESs is a logical segue to the next chapter, in which we more generally analyze trends in military experience among members of the AW.

CHAPTER SIX

The Military Acquisition Workforce and Its Implications for the Civilian Acquisition Workforce

In this chapter we present insights on the military AW based on our analysis of DoD data. The first section provides a descriptive overview of the military AW. The second section describes the relationship between the military and civilian workforces and how it varies by service. The third section discusses the military AW as a source of new hires into the civilian AW.

Descriptive Overview

In parallel to the civilian AW, there are also members of the military coded as being in the AW. Figure 6.1 shows the time trajectory of the number of military members in the AW. Figure 6.1 and other data presented in this section are from the WEX. The military AW total is roughly an order of magnitude less than the total in the civilian AW.

As shown in Figure 6.2, military AW personnel are predominantly in the Air Force.

Figure 6.1
Military Members in the Acquisition Workforce, September 30 Annual Snapshots

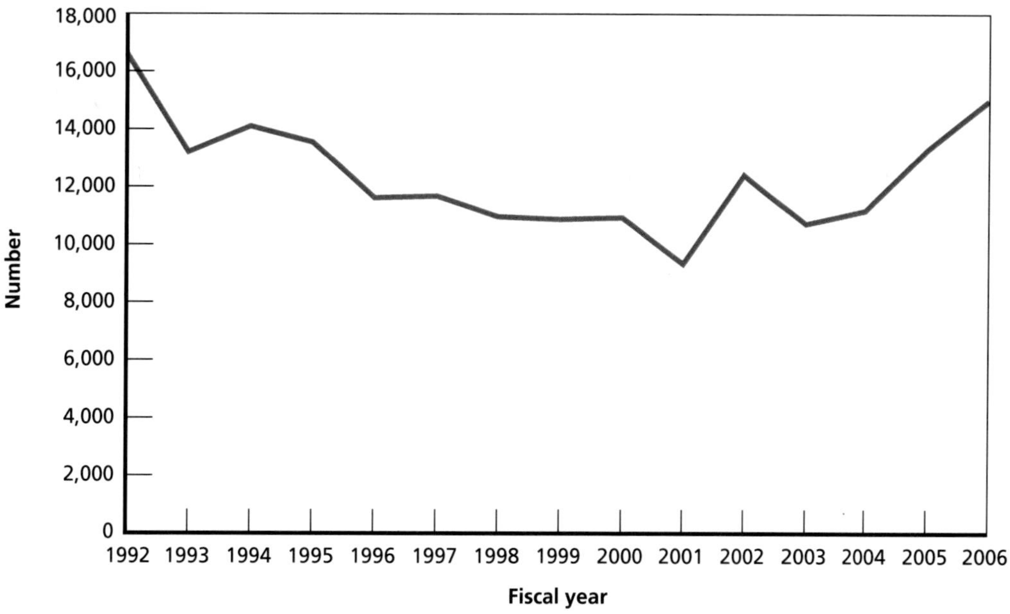

RAND *TR572-6.1*

Figure 6.2
AW Military Inventory, by Service, 2006

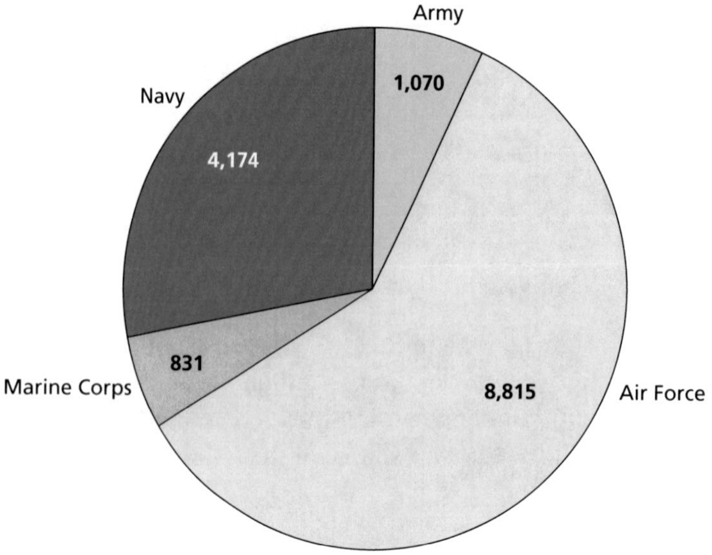

RAND *TR572-6.2*

The Military Acquisition Workforce as a Share of the Total Acquisition Workforce

The previous section indicates that, while the AW includes both military and civilian personnel, military personnel are a distinct minority. The share of a service's workforce represented by military personnel varies by service, suggesting that services make different choices regarding how and when to use military personnel in acquisition roles.

Figure 6.3 displays civilian and military totals by service. Consistent with Figure 6.2, military personnel are most prominent in the Air Force's acquisition workforce.

Figure 6.4 presents a scatter plot in which the x-axis is the percentage of the service's civilians in the AW and the y-axis is the percentage of the service's military in the AW.

Nearly one in five Army civilians are in the AW, but fewer than one in 1,000 Army military are in the AW. The Air Force has the greatest percentage of its military in the AW, but even its total is only about 1.4 percent.[1]

As shown in Figure 6.5, about 85 percent of military personnel in the AW are officers, in near-symmetrical contrast to the less than 15 percent of military personnel, overall, who are officers.

[1] These figures reflect the share of military in the AW relative to the total force. However, military members performing acquisition functions are typically not part of the "expeditionary" or deployable segment of the military population, but are part of the "institutional" or support segment. The fraction of military in the AW relative to the number in the institutional arm of the service would be higher.

Figure 6.3
Total Civilian and Military Acquisition Workers, by Service, 2006

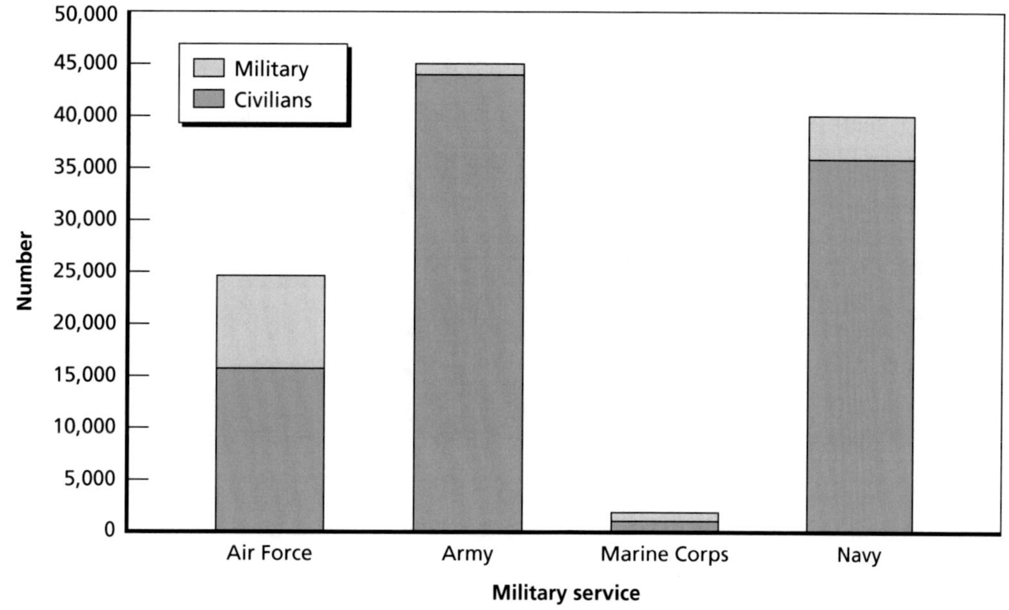

RAND *TR572-6.3*

Figure 6.4
Percentages of Civilians and Military Personnel in the Acquisition Workforce, 2006

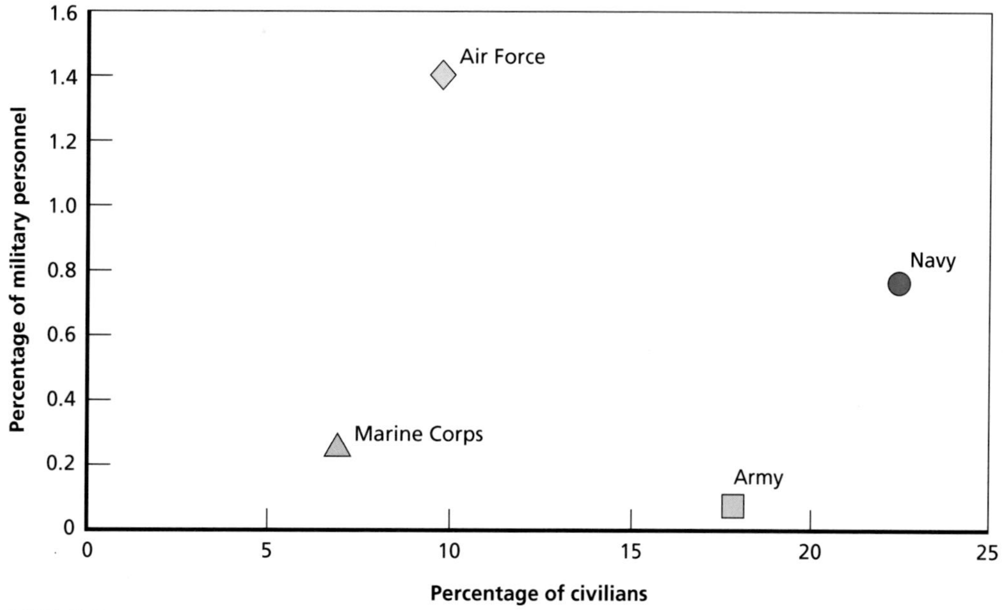

RAND *TR572-6.4*

Figure 6.5
Military Personnel, by AW and Enlistment Status, 2006

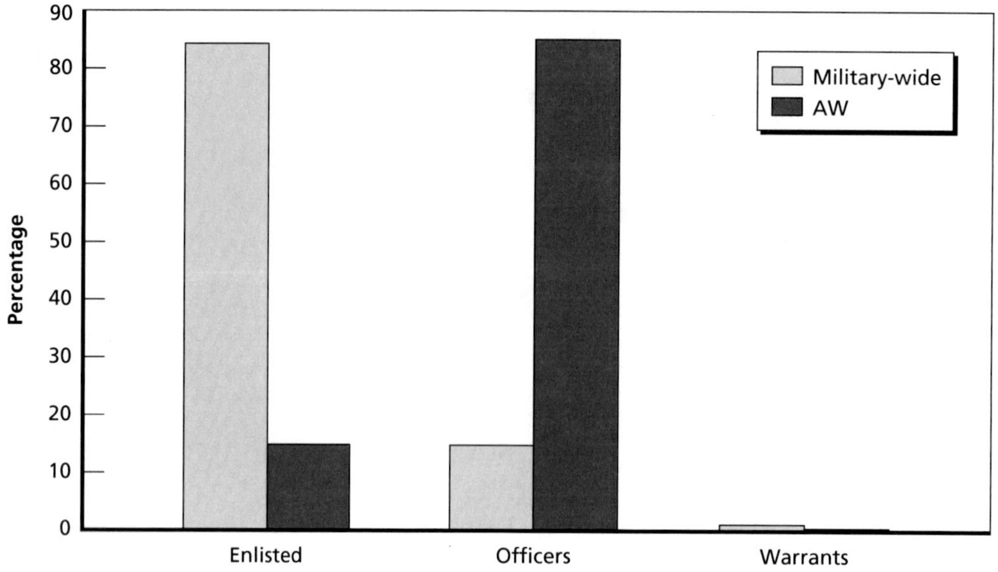

RAND *TR572-6.5*

In Chapter Three, we presented information on the career field distribution of the civilian AW (Figure 3.5). We noted that a majority of civilians are in the career fields of Systems Planning, R&D, and Engineering and Contracting and that only 7 percent of civilian acquisition employees are in Program Management. In Figure 6.6 we present information on the acquisition career field for the military AW in FY 2006. Although Contracting and Systems Planning, R&D, and Engineering are important career fields for the military acquisition workforce, the largest share of military acquisition workers are in the Program Management career field (29 percent of the total). An analysis of the data underlying Figures 3.5 and 6.6 (not presented here) illustrates areas in which military members are either over- or underrepresented. Whereas the military workforce constitutes only 12 percent of the total acquisition workforce, it makes up about a third of the Program Management community. Military members are also overrepresented in the Test and Evaluation Engineering career field, where they make up one-quarter of all employees. In contrast, there are a number of career fields with few or no military personnel: Auditing, Science/Technology, Manufacturing and Production, Purchasing and Procurement, and Industrial Property Management.

The Military Acquisition Workforce as a Source of New Hires into the Civilian Acquisition Workforce

Military members of the acquisition workforce are of particular interest to us because we suspect they are prominent candidates to leave the military and become leaders of the civilian AW.

We cannot say with any degree of confidence what fraction of current AW civilian employees have prior military experience. Military experience flags in the civilian inventory file appear to be unreliable. As noted previously, our WEX data go back only to 1975. We know that numerous

Figure 6.6
Career Field Distribution for the Military AW, FY 2006

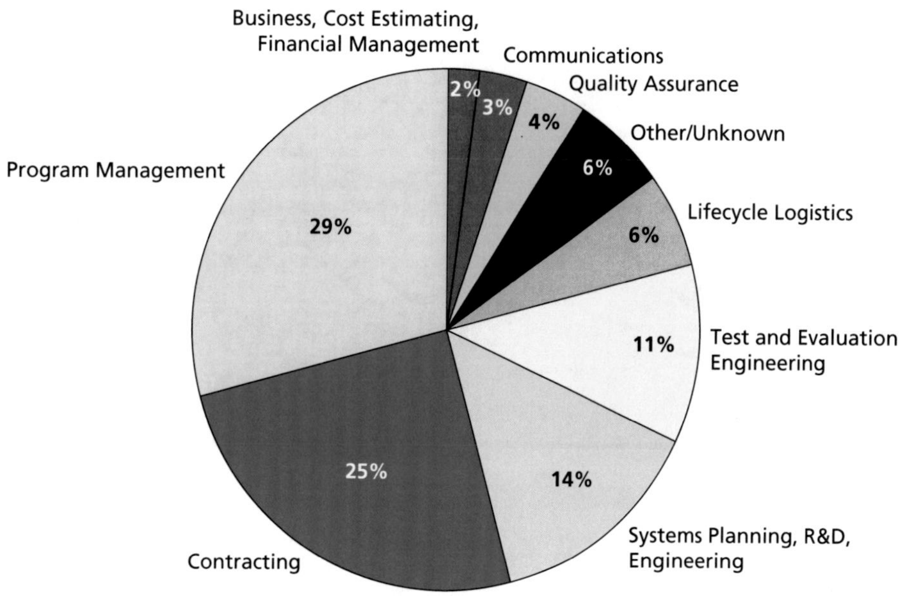

RAND *TR572-6.6*

senior DoD civilian employees served in the military during the Vietnam era or earlier, but we have no confidence in our ability to identify those individuals using our data.

We have greater confidence, however, in our ability to calibrate the level of military experience among new civilian hires. As discussed in Chapter Two, the only new hires whose military experience we could not tabulate would be those who left the military before 1975. Although there are doubtlessly some, we do not think there are many new civilian employees hired in 1992 or later whose military service ended before 1975.

As shown in Figure 6.7, there has been a marked and interesting trend in both DoD-wide and AW rates of military experience of new civilian hires.

In FY 2006, over one-third (39.6 percent) of new AW and over half (54.4 percent) of new DoD civilian employees had prior military experience. Both these rates are higher than new hire military-experience rates in these categories seen in any fiscal year dating back to 1993. The AW has consistently hired fewer civilians with military experience than has the DoD as a whole.

We do not yet have a complete explanation for this markedly upward trend. Military-civilian conversions may have encouraged hiring civilians with military experience—often the very individuals who previously performed the function as members of the military. Also, as shown in Figure 3.4, there is a comparative dearth of DoD civilian employees with 5–14 YOS, a consequence of the 1990s drawdown. New hires with military experience may be serving as a replacement for unavailable mid-career civilians.

Another factor may be an October 1999 policy change that removed an income cap that had discouraged higher-ranking military retirees from taking federal jobs. Prior to October 1, 1999, there was a pay cap that limited the combined total of federal civilian basic salary plus military retired pay to the Executive Level V compensation level (which was $110,700 at the time).

Figure 6.7
Percentage of New Civilian Hires with Prior Military Experience (1975 or later)

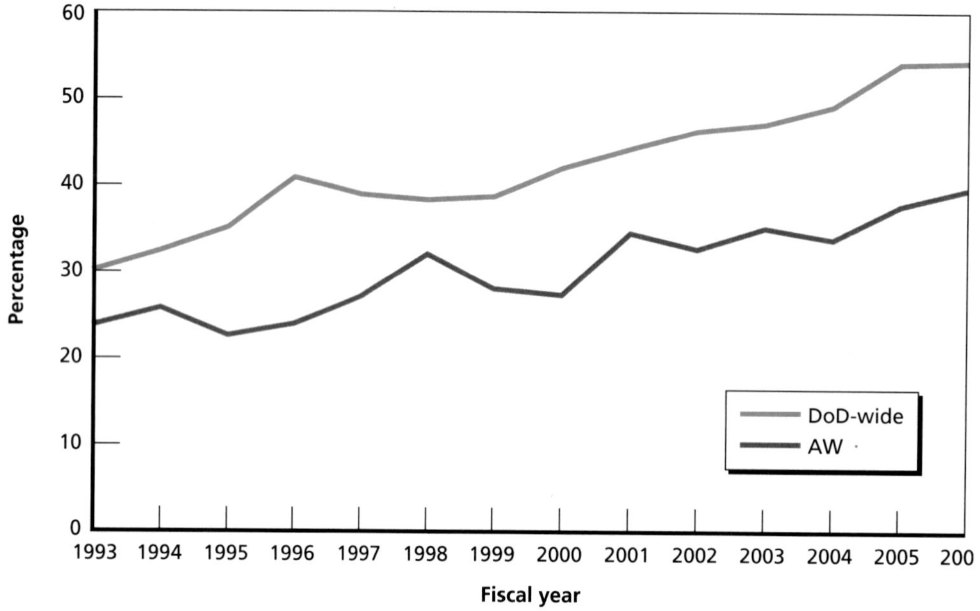

RAND *TR572-6.7*

We found mixed evidence that removal of this pay cap contributed to the increase in the number of new hires with military experience. We expected that removal of the pay cap would be most relevant for military retirees drawing pensions. Unfortunately, our data do not clearly identify military retirees, but the WEX does tell us the new hire's last rank as a military member. We defined a military new hire to be *high-ranking* if he or she attained a rank of E7 or above, O5 or above, or WO3 or above while in the military. These are the ranks that we think would typically correspond to full-career military retirees (though, of course, there will always be exceptions, e.g., prior enlisted personnel who later become officers and retire as O4s).

In Figure 6.8, we plot the proportion of those new hires with military experience who were high-ranking as we define it. In the AW, the proportion who were high-ranking has grown in recent years, consistent with the theory that the pay cap had been a binding constraint, but we do not observe any change in the percentage of high-ranking new hires DoD-wide (although, as shown in Figure 6.7, there was a marked increase in the proportion of new hires with military experience both in the AW and DoD-wide).

We conclude from Figure 6.8 that the October 1999 removal of the pay cap cannot be the sole explanation for the marked increase in the proportion of new hires with military experience, although it appears to have had the expected effect on the AW.

Echoing Figure 5.6 concerning the Senior Executive Service members, military services tend to hire their own service's veterans. Figure 6.9 shows FY 2006 AW hires with prior military experience and the percentage of those hires who came from the respective services.

Although military members represent a minority of the AW overall, they appear to be an important and growing source of future AW civilian leaders, especially in light of the undersized (drawdown-era) cohorts of DoD civilians.

Figure 6.8
Percentage of New Civilian Hires with Military Experience Who Were High-Ranking

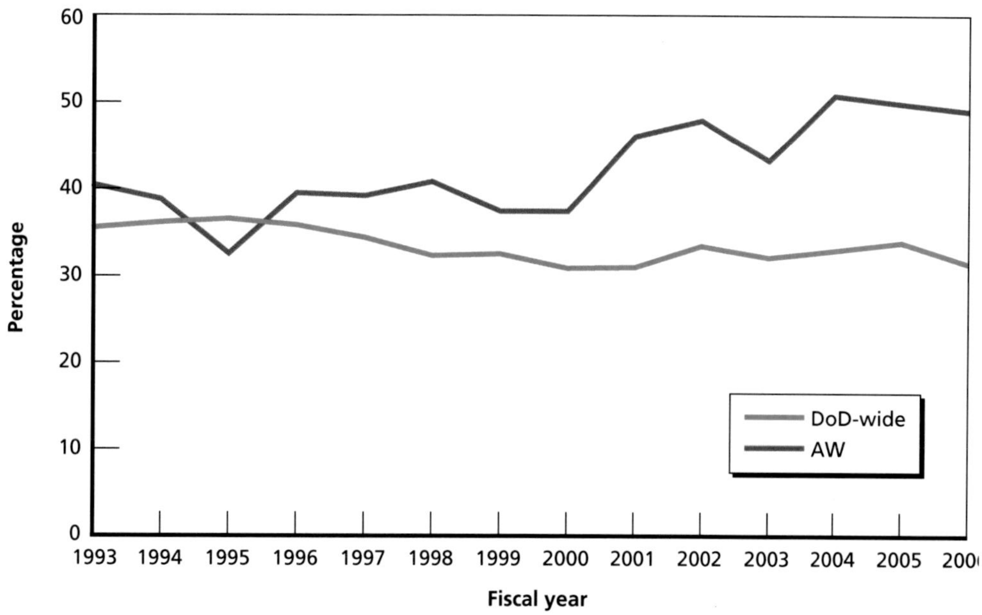

RAND *TR572-6.8*

Figure 6.9
Military Service of New Civilian Acquisition Workforce Hires, by Civilian Hiring Service, 2006

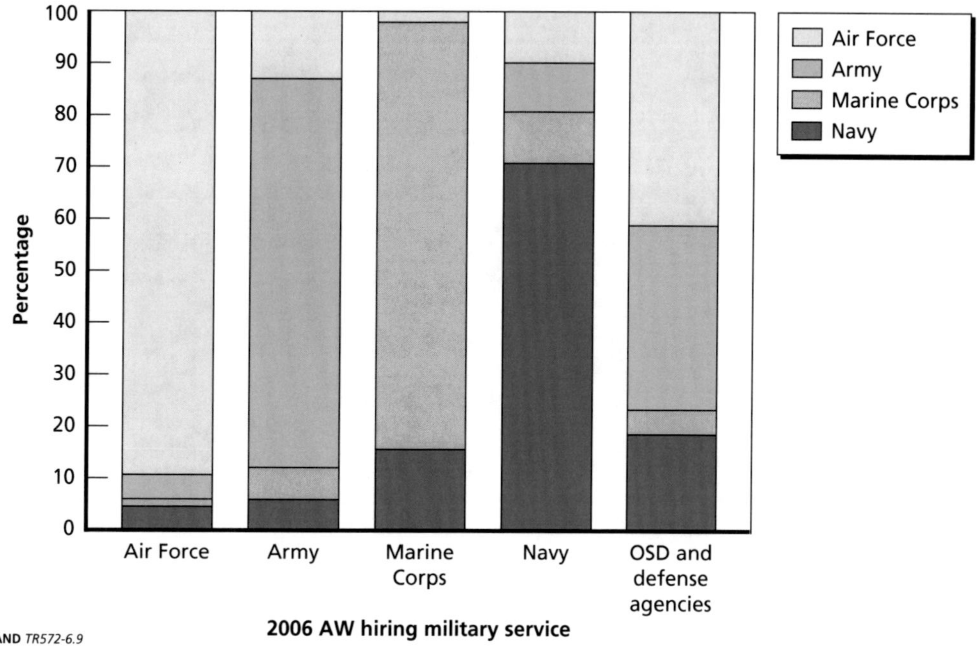

RAND *TR572-6.9*

CHAPTER SEVEN

Conclusions

The DoD slice of the federal acquisition workforce is a large workforce in itself, consisting of more than 126,000 military and civilians in over a dozen different services and agencies and representing several functional communities. In turn, much of the strategic human capital planning in DoD occurs within individual services and agencies. Still, DoD-wide visibility over such groups is essential to effective management and planning for a critical workforce that cuts across organizational boundaries. Through an analysis of the acquisition workforce (both current snapshots and trends over time), we have assisted AT&L HCI/DAU in identifying and better understanding analytical challenges to acquisition workforce management and developing strategies for improving strategic workforce planning in the future.

Findings

AW civilians differ from "regular" civilian employees of the DoD. AW civilians are, on average, better educated and more senior. They also have lower attrition, even controlling for education and years of service. Indeed, AW civilians appear unusually attached to their jobs. To the extent that one may be concerned about the loss of experienced AW civilian employees, the fact that this population has attrited at below-normal rates is cause for optimism.

Membership in the AW has not been a stable construct, however. There have been large-scale, service-level recategorizations into and out of the AW without commensurate apparent changes in what workers have been doing. We suspect that individuals have been categorized in and out of the AW without their jobs changing and without the worker even realizing that anything has changed. Administrative recategorizations pose challenges for strategic human capital planning. As noted in Chapter One, various definitions of the acquisition workforce have been used at various points in time. This may have led to some confusion on the part of those responsible for categorizing workers as part of the AW. In addition, the flexibility of the definition allows for organizational variation in its interpretation. This, combined with political pressures to reduce the size of the acquisition workforce that have emerged from time to time, may contribute to administrative recategorizations.

As of September 30, 2006, there were 454 SESs in the AW. Not surprisingly, the vast majority of these SESs were long-time DoD civilian employees. DoD appears to value depth over breadth in promotion to SES status—very few of these SESs had worked outside their current military service since 1992. In terms of the relative handful of AW SESs who were comparatively new to DoD civilian employment, several had previously been high-ranking military officers.

To the extent the AW hires new civilian employees, a growing fraction of them in recent years comprises military veterans. This matches a DoD-wide trend toward hiring more civilians who have previously served in the military.

It is not clear to us that there is anything unique—or dire—about the civilian workforce issues and challenges facing the acquisition workforce right now. The AW is more senior and hence more retirement-vulnerable than is typical DoD-wide, but members of the AW have consistently exhibited strong attachment to DoD employment. Indeed, the AW would face much greater challenges if its civilian workers started to attrit in the way "regular" DoD civilian employees do.

This could happen at some point, however, if the acquisition environment changes significantly or if other civilian personnel issues, such as the introduction of the National Security Personnel System, produce a negative effect. In any case, the attention given to AW human capital issues is, to a considerable extent, a tribute to those workers' importance.

Recommendations

Better Definition and Tracking of the Acquisition Workforce Would Improve Workforce Planning

Our analyses reveal that the definition of the acquisition workforce is a fluid concept, with numerous DoD civilians being recategorized into and out of the acquisition workforce each year. A lack of clarity as to the definition of the workforce and how it varies across organizations is a barrier to effective management of the AW and the development of recruitment, training, and retention policies. OSD should work together with the services to revise the data collection policy guidance in order to improve consistency across organizations.

More-Detailed Analysis of the Current Acquisition Workforce and Historical Trends Could Yield Additional Insight

In this report, we illustrate the types of supply analysis that can be supported by existing data, and we highlight some examples of analyses that could be informative for managers. While much of the workforce analysis should occur at the service or major command level, there are some cross-cutting areas in which a DoD-wide perspective, including a comparison across services and agencies, might be particularly valuable. For example, a better understanding of the careers and performance of individuals in key leadership positions and the careers of acquisition workforce alums might be particularly useful for both DoD and the services. A retrospective analysis of individuals who are currently filling civilian key leadership positions could examine the extent to which these individuals came from the military, from special intern programs, or from other career development programs. It could also consider the geographic and organizational mobility of individuals holding key leadership positions, as well as their breadth of experience. An analysis of AW alums (i.e., current DoD employees who are not currently part of the AW but have been at some point in the past) would provide an estimate of individuals who are currently part of the DoD workforce, have some acquisition experience, but are not currently part of the acquisition workforce. This population could be considered a "reserve pool" of acquisition workers. An analysis of these individuals could characterize where they are located, how long they were part of the acquisition workforce before leaving, whether they were certified, and how many years of service of service they had.

A Better Understanding of the Post-Military Careers of the Military Acquisition Workforce Could Be Useful to Acquisition Workforce Managers

As we have shown, a growing fraction of new hires into DoD's civilian AW have prior military experience. The military workforce is thus an important hiring pool for the civilian workforce. A deeper understanding of the career opportunities and choices made by members of the military AW upon separation from active duty would help DoD better understand the challenge it faces in recruiting and retaining a high-quality civilian workforce. In addition, an improved understanding of the careers of new civilian hires with prior military experience would improve DoD's civilian AW management.

Workforce Analysis Is Only One Step in an Overall Strategic Human Capital Planning Effort

In this report, we have provided an overview of the DoD acquisition workforce and some concrete examples of the types of workforce supply analyses that can be supported by DoD data. Nevertheless, it is important to note that supply analysis is only one (and arguably not the most important) step in strategic human capital planning. Such analyses must be combined with demand analyses. A more systematic and data-based understanding of workload drivers for the AW and the relationship between changes in the acquisition process and workload levels would facilitate strategic human capital planning for the AW.

Better Information on the Contractor Workforce in Acquisition Functions Is Needed

Even with perfect information on workforce demand, it will be impossible for workforce managers to conduct effective gap analysis and develop appropriate workforce strategies given the lack of information on contractors who are performing acquisition-related functions. The DoD Inspector General has emphasized the difficulty involved in obtaining even the most basic information, such as the number of contractors employed in acquisition-related functions, not to mention more-detailed information on their area of specialization and their certification levels. Such information is critical for managers interested in assessing the health of the acquisition workforce.

References

Congressional Research Service Report (1999). *Defense Acquisition Workforce: Issues for Congress.* March 11, 1999.

Defense Acquisition University (2007). *Defense Acquisition Structures and Capabilities Review—Report, Fiscal Year 2006.* Fort Belvoir, Va.

DoD—See U.S. Department of Defense.

DoD Instruction 5000.55—*See* U.S. Department of Defense (1991).

DoD Instruction 5000.66—*See* U.S. Department of Defense (2005).

Executive Office of the President, Office of Management and Budget (2002). *The President's Management Agenda, Fiscal Year 2002.* As of May 9, 2008:
http://www.whitehouse.gov/omb/budget/fy2002/mgmt.pdf

Fairhall, James (1987). "The Case for the $435 Hammer—Investigation of Pentagon's Procurement." *Washington Monthly*, January 1987.

Garcia, A., H. Keyner, T. Robillard, and M. Van Mullekom (1997). "The Defense Acquisition Workforce Improvement Act: Five Years Later." *Acquisition Review Quarterly*, 4(3): 295.

Gates, Susan M., Christine Eibner, and Edward G. Keating (2006). *Civilian Workforce Planning in the Department of Defense: Different Levels, Different Roles.* Santa Monica, Calif.: RAND Corporation, MG-449-OSD. As of May 5, 2008:
http://www.rand.org/pubs/monographs/MG449/

Hanks, Christopher H., Elliot I. Axelband, Shuna Lindsay, Mohammed Rehan Malik, and Brett D. Steele (2005). *Reexamining Military Acquisition Reform: Are We There Yet?* Santa Monica, Calif.: RAND Corporation, MG-291-A. As of May 5, 2005:
http://www.rand.org/pubs/monographs/MG291/

Schirmer, Peter, Harry J. Thie, Margaret C. Harrell, and Michael S. Tseng (2006). *Challenging Time in DOPMA: Flexibility and Contemporary Military Officer Management.* Santa Monica, Calif.: RAND Corporation, MG-451-OSD. As of May 5, 2005:
http://www.rand.org/pubs/monographs/MG451/

U.S. Department of Defense (1991). DoD Instruction 5000.55, "Reporting Management Information on DoD Military and Civilian Acquisition Personnel and Position." As of May 9, 2008:
http://www.dtic.mil/whs/directives/corres/html/500055.htm

——— (2005). DoD Instruction 5000.66, "Operation of the Defense Acquisition, Technology, and Logistics Workforce Education, Training, and Career Development Program." December 21, 2005. As of May 9, 2008:
http://www.dtic.mil/whs/directives/corres/pdf/500066p.pdf

——— (2006). *Quadrennial Defense Review Report.* February 6, 2006. As of May 5, 2008:
http://www.defenselink.mil/pubs/pdfs/QDR20060203.pdf

U.S. Department of Defense, Acquisition, Technology, and Logistics (2006). *AT&L Human Capital Strategic Plan,* Volume 1. As of May 5, 2008:
http://www.dau.mil/workforce/hcsp.pdf

——— (2007). *AT&L Human Capital Strategic Plan, v 3.0.* As of May 20, 2008: http://www.dau.mil/workforce/hcsp.pdf

U.S. Department of Defense, Office of the Inspector General (2006). *Human Capital: Report on the DoD Acquisition Workforce Count.* D-2006-073.

U.S. Government Accountability Office (2002). *Acquisition Workforce: Department of Defense's Plans to Address Workforce Size and Structure Challenges.* GAO-02-630.

U.S. Office of Personnel Management (1997). *FERS—Federal Employees Retirement System Transfer Handbook: A Guide to Making Your Decision.* RI 90-3. As of May 7, 2008: http://www.opm.gov/retire/fers_election/fersh/hb.pdf

Vernez, Georges, Albert A. Robbert, Hugh G. Massey, and Kevin Driscoll (2007). *Workforce Planning and Development Processes: A Practical Guide.* Santa Monica, Calif.: RAND Corporation, TR-408-AF. As of May 5, 2008: http://www.rand.org/pubs/technical_reports/TR408/